T0252528

MARITIME BOUNDARIES

The global political map is undergoing a process of rapid change as former states disintegrate and new states emerge. At sea, boundary delimitation between coastal states is continuing unabated. These changes could pose a threat to world peace if they are not wisely negotiated and carefully managed.

Maritime Boundaries presents a variety of cases illustrating the implications of recent approaches to maritime territorial jurisdictions.

World Boundaries is a unique series embracing the theory and practice of boundary delimitation and management, boundary disputes and conflict resolution, and territorial change in the new world order. Each of the five volumes – *Global Boundaries, The Middle East and North Africa, Eurasia, The Americas* and *Maritime Boundaries* – is clearly illustrated with maps and diagrams and contains regional case studies to support thematic chapters. This series will lead to a better understanding of the means available for the patient negotiation and peaceful management of international boundaries.

Gerald H. Blake is Director of the International Boundaries Research Unit at the University of Durham.

WORLD BOUNDARIES SERIES
Edited by Gerald H. Blake
Director of the International Boundaries Research Unit
at the University of Durham

The titles in this series are:

GLOBAL BOUNDARIES
Edited by Clive H. Schofield

THE MIDDLE EAST and NORTH AFRICA
Edited by Clive H. Schofield and Richard Schofield

EURASIA
Edited by Carl Grundy-Warr

THE AMERICAS
Edited by Pascal Girot

MARITIME BOUNDARIES
Edited by Gerald H. Blake

MARITIME BOUNDARIES

World Boundaries volume 5

Edited by Gerald H. Blake

London and New York

First published 1994
by Routledge
2 Park Square, Milton Park, Abingdon, Oxon, OX14 4RN

Simultaneously published in the USA and Canada
by Routledge
270 Madison Ave, New York NY 10016

Reprinted 1997

Transferred to Digital Printing 2007

© 1994 International Boundaries Research Unit

Typeset in 10pt September by Solidus (Bristol) Limited

British Library Cataloguing in Publication Data
A catalogue record for this book is available from the British Library

Library of Congress Cataloging in Publication Data
Maritime boundaries / edited by Gerald H. Blake.
p. cm. – (World boundaries ; v. 5)
Includes bibliographical references and index.
1. Territorial waters. 2. Economic zones (Maritime law).
I. Blake, Gerald Henry. II. Series.
JX4131.M293 1994
320.1'2–dc20 93-35761

ISBN 0–415–08835–6
5-vol. set: ISBN 0–415–08840–2

Publisher's Note
The publisher has gone to great lengths to ensure the quality of this
reprint but points out that some imperfections in the original
may be apparent

CONTENTS

CONTENTS

LIST OF ILLUSTRATIONS

Figures

Tables

Appendices

NOTES ON CONTRIBUTORS

Peter B. Beazley is former Territorial Waters Officer for the United Kingdom and was the 1991 IBRU Visiting Fellow at the University of Durham.

Rodman R. Bundy is a partner in the Frere Cholmeley Conseils Juridiques, Paris.

Galo Carrera is Director of Geometrix, Dartmouth, N.S., Canada.

Jonathan I. Charney is Professor of Law at Vanderbilt University, Nashville, Tennessee.

Douglas Day is Professor of Geography at Saint Mary's University, Halifax, N.S. Canada.

Giampiero Francalanci works at the International Law Institute at the University of Parma.

David Freestone is Professor of Law and associate of the Institute of Estuarine and Coastal Studies at the University of Hull.

Charles E. Harrington works at the National Ocean Service, Nautical Charting Division, Rockville, Maryland, USA.

Geoffrey Marston is a Fellow of Sidney Sussex College, Cambridge.

John Pethick is Director of the Institute of Estuarine and Coastal Studies at the University of Hull.

Tullio Scovazzi is a professor at the International Law Institute at the University of Parma, Italy.

Robert W. Smith is in the Department of Environmental and Ocean Affairs, US Department of State, Washington.

FOREWORD

The period since 1945 has seen a growing interest in the practical aspects of maritime boundaries. Coastal state practice, impelled by economic and strategic considerations, has encroached upon maritime space once regarded as being outside territorial jurisdiction. The great revision of the law of the sea which culminated in the Montego Bay Convention of 1982 has endorsed the right of coastal states to maritime zones extending up to 200 nautical miles, and in respect of the continental shelf in some circumstances even further. It is not surprising therefore that the risk of dispute and even conflict over such expansionism has increased in proportion to the growth in the actual or potential interface between states' maritime claims. To give one example: the entire waters and subsoil of the Mediterranean are potentially under the jurisdiction of one or another of its coastal states; if each were to exercise its entitlement, a substantial number of disputes would thereby be generated, not so much over the entitlement itself as over the delimitation of the zones thus claimed. Meanwhile, settlement of maritime boundary disputes is proceeding in an *ad hoc* and substantially bilateral manner in the form of contentious cases brought before the International Court of Justice and arbitral tribunals, and also in the form of treaties. The justification for the treatment of maritime boundaries in a separate volume of the proceedings of the Second Durham IBRU Conference in 1991 is therefore manifest.

Turning to the contents of Volume 5 in the series, a quadripartite distinction may be made. First, the paper by Jonathan Charney outlines the preparation of what is an essential tool in the hands of those concerned with maritime delimitation the world over, whether as government officials, legal advisers, technical experts or academic writers. The publication of the Maritime Boundary Project of the American Society of International Law is a highly significant event.

x

The paper by Rodman Bundy has elements, too, of a global overview, though its subject-matter is narrower in that it concentrates on state practice in maritime delimitations. Bundy also examines that most difficult question, the relevance of state practice as a material element in the formation of rules of customary international law on this subject. A trio of papers, written by experts, concerns practical aspects of maritime boundaries. The need for a wider knowledge of such technical material is undoubted. Charles Harrington describes how one state, the USA, has indicated maritime limits on its official maps. Another way in which such knowledge may be made available for the benefit of 'end-users' is described in the paper on the Delmar System by Galo Carrera. Peter Beazley's paper on drying coral reefs as basepoints is a critique by a hydrographer of lawyers' words now embodied in the 1982 Montego Bay Convention.

A third group of papers falls into the category of localized 'case-studies'. The paper by Robert Smith on the USA/Soviet Union treaty of 1990 delimiting the North Pacific and Arctic Oceans is not only authoritative in that the author was a member of the US negotiating team, but its subject matter will have gained some element of additional significance, and perhaps controversy, in the light of the subsequent disintegration of the Soviet Union and the assumed continuation of its legal personality in the form of the Russian Federation. Douglas Day's paper on the management of transboundary fish stocks on both sides of the North Atlantic looks through the eyes of a resource geographer at a matter which has created considerable acrimony in the last decade. Whether the arbitral award in June 1992 delimiting as between France and Canada the maritime zones of the French archipelago of St Pierre-Miquelon will abate that element of the problem remains to be seen. The paper by Giampiero Francalanci and Tullio Scovazzi on Egyptian straight baseline legislation is a case-study of a general question vital in maritime legislation, namely the ascertainment of the baseline from which all maritime zones are measured.

Finally, the concept of boundary stability discussed in Geoffrey Marston's paper could have been included equally in one of the other volumes as it deals with an issue relevant to both land and maritime boundary delimitations. Although relatively tranquil in its application to the maritime zone, it is currently undergoing severe turbulence in the wake of political disintegration within the areas of what used to be Yugoslavia and the Soviet Union.

Geoffrey Marston

SERIES PREFACE

The International Boundaries Research Unit (IBRU) was founded at the University of Durham in January 1989, initially funded by the generosity of Archive Research Ltd of Farnham Common. The aims of the unit are the collection, analysis and documentation of information on international land and maritime boundaries to enhance the means available for the peaceful resolution of conflict and international transboundary cooperation. IBRU is currently creating a database on international boundaries with a major grant from the Leverhulme Trust. The unit publishes a quarterly *Boundary and Security Bulletin* and a series of *Boundary and Territory Briefings*.

IBRU's first international conference was held in Durham from September 14–17 1989 under the title of 'International Boundaries and Boundary Conflict Resolution'. The 1989 Conference proceedings were published by IBRU in 1990 under the title *International Boundaries and Boundary Conflict Resolution* edited by C.E.R. Grundy-Warr. The theme chosen for our second conference was 'International boundaries: fresh perspectives'. The aim was to gather together international boundary specialists from a variety of disciplines and backgrounds to examine the rapidly changing political map of the world, new technical and methodological approaches to boundary delimitations, and fresh legal perspectives. Over 130 people attended the conference from 30 states. The papers presented comprise four of the five volumes in this series (Volumes 1–3 and Volume 5). Volume 4 largely comprises proceedings of the Second International Conference on boundaries in Ibero–America held at San José, Costa Rica, November 14–17 1990. These papers, many of which have been translated from Spanish, seemed to complement the IBRU Conference papers so well that it was decided to ask Dr Pascal Girot, who is coordinator of a major project on border regions in Central America based at CSUCA (The Confederation of Central American Universities) to edit them for inclusion in the series. Volume 4 is thus symbolic of the practical cooperation which IBRU is developing with a number of institutions overseas whose objectives are much the same as IBRU's. The titles in the *International Boundaries* series are:

Volume 1 *Global Boundaries*
Volume 2 *The Middle East and North Africa*
Volume 3 *Eurasia*
Volume 4 *The Americas*
Volume 5 *Maritime Boundaries*

The papers presented at the 1991 IBRU conference in Durham were not specifically commissioned with a five-volume series in mind. The papers have been arranged in this way for the convenience of those who are most concerned with specific regions or themes in international boundary studies. Nevertheless the editors wish to stress the importance of seeing the collection of papers as a whole. Together they demonstrate the ongoing importance of research into international boundaries on land and sea, how they are delimited, how they can be made to function peacefully, and perhaps, above all, how they change through time. If there is a single message from this impressive collection of papers it is that boundary and territorial changes are to be expected, and that there are many ways of managing these changes without resort to violence. Gatherings of specialists such as those at Durham in July 1991 and at San José in November 1990 can contribute much to our understanding of the means available for the peaceful management of international boundaries. We commend these volumes as being worthy of serious attention not just by the growing international community of border scholars, but by decision-makers who have the power to choose between patient negotiation and conflict over questions of territorial delimitation.

Gerald H. Blake
Director of IBRU
Durham, January 1993

ACKNOWLEDGEMENTS

Much of the initial work on these proceedings was undertaken by IBRU's executive officer Carl Grundy-Warr before his appointment to the National University of Singapore early in 1992. It has taken a team of editors to complete the task he began so well. Elizabeth Pearson and Margaret Bell assisted in the preparation of the manuscripts for several of these volumes, and we acknowledge their considerable contribution. John Dewdney came to our rescue when difficult editorial work had to be done. In addition, many people assisted with the organisation of the 1991 conference, especially my colleagues Carl Grundy-Warr, Greg Englefield, Clive Schofield, Ewan Anderson, William Hildesley, Michael Ridge, Chng Kin Noi and Yongqiang Zong. Their hard work is gratefully acknowledged. We are most grateful to Tristan Palmer and his colleagues at Routledge for their patience and assistance in publishing these proceedings, and to Arthur Corner and his colleagues in the Cartography Unit, Department of Geography, University of Durham for redrawing most of the maps.

Gerald H. Blake
Director, IBRU

1

THE AMERICAN SOCIETY OF INTERNATIONAL LAW MARITIME BOUNDARY PROJECT

Jonathan I. Charney

INTRODUCTION

I have been invited to report on the results and status of the American Society of International Law Maritime Boundary Project which I directed with the assistance of Lewis M. Alexander.[1] Over twenty other leading experts from around the world participated. The results of the research are contained in a typescript manuscript of approximately 3,500 pages, which will be published in two volumes by Martinus Nijhoff Publishers early in 1992 under the title *International Maritime Boundaries* and edited by Charney and Alexander.[2] The book will contain the texts of the 137 known maritime boundary settlements (in English),[3] maps illustrating each settled boundary, ten small-scale maps illustrating all of the settled boundaries in different regions of the world, written analyses of each settled boundary, ten papers analyzing regional practices and circumstances, nine papers analyzing different aspects of maritime boundary delimitation practices, and an introduction containing an overview of the book.

The papers published in the book benefited from substantial collegial exchanges; however, the views expressed in the papers are those of the individual authors and do not necessarily reflect the views of organizations or governments with which the authors may be associated. Consequently, there are no conclusions adopted by the project *per se* to be reported.

THE CONTEXT AND OBJECTIVES

While states have delimited their maritime boundaries for centuries, so much has changed in recent years that the most relevant data postdates

1

1940. Since that year coastal states have settled 137 maritime boundaries, although many more remain unresolved. At least a dozen disputes relating to maritime boundaries have been submitted for resolution to international courts, conciliators, and arbitrators. During the last eight years, there have been, on average, four or five settlements each year, with many others under active negotiation or study.

The product of the 1958 Law of the Sea Conference was the relatively indeterminate equidistance-special circumstances rule contained within the Convention on the Territorial Sea and the Contiguous Zone, and the Convention on the Continental Shelf. The matter was subsequently taken up by the International Court of Justice in the *North Sea Continental Shelf Cases*. The Court did provide some guidance to the parties, but pronounced a multifactored rule that may have made the law more indeterminate. The 1982 Convention on the Law of the Sea, which was the product of the Third United Nations Conference on the Law of the Sea, retained the less-than-determinate equidistance-special circumstances language for maritime boundaries in the territorial sea, but jettisoned that language for boundaries in the continental shelf and the exclusive economic zone. This new language spoke in terms of 'equitable solutions', dropping all references to equidistance and special circumstances. A reference to international law provides an ambiguous connection to the old language and customary international law. Litigation and arbitration have produced equally indeterminate results with respect to the operative norm. From the very beginning the codified law and third-party decisions have stressed the primary objective that states should settle their maritime boundaries by agreement. Only in the absence of agreement would positive law dictate the result in any particular case.

While the international community has sought to identify the norms applicable to maritime boundary delimitations, courts and tribunals charged with addressing specific disputes have not given much attention to the state practice. This is despite the fact that settlements of maritime boundaries by agreement may contribute to the evolution of the relevant positive norm of international law.

It was the purpose of this project to study existing maritime boundary settlements to discover what light they might shed on the rules and practices relevant to the resolution of maritime boundary disputes. The project was designed to study each of the known boundaries in a systematic way in order to compare the approaches used to resolve these disputes.

THE WORK OF THE REGIONAL EXPERTS

At the first stage, the world was divided into ten regions. Experts on those regions were commissioned to study the maritime boundaries in the region for which they assumed primary responsibility. They prepared individual reports on each boundary for the purpose of identifying those considerations that entered into the boundary settlements. The 'Regional Experts' who participated in the first stage of the project were:

1 North America – Lewis M. Alexander[4]
2 Middle America and Caribbean – Kaldone Nweihed
3 South America – Eduardo Jiménez de Aréchaga
4 Africa – Andronico O. Adede
5 Central Pacific/East Asia – Choon-Ho Park[5]
6 Indian Ocean – J.R. Victor Prescott
7 Persian Gulf – Robert F. Pietrowski Jr. and Lewis M. Alexander[6]
8 Mediterranean and Black Seas – Tullio Scovazzi[7]
9 Northern and Western Europe – David Anderson
10 Baltic Sea – Erik Franck

The Regional Experts were particularly asked to explore nine categories of considerations that might have played a role in the delimitations. Those categories are as follows:

Political, strategic, and historical considerations

If and how political factors, security interests, historical considerations, contemporary usages, or perceived legal constraints affected the negotiation or location of the boundary, e.g. geopolitical power, military power, strategic considerations, contemporary relations between the parties, conflict avoidance, other maritime boundaries or disputes, navigation and overflight interests, historical claims, established resource exploitation activities, and contemporaneous accommodation of related or unrelated interests of the parties.

Legal regime considerations

What is the juridical character of the areas delimited by the boundary agreement? Are there provisions for dispute settlement and cooperation? If and how the movement from a continental shelf to an exclusive economic or fisheries zone boundary (including grey zones) affected the location of the maritime boundary.

Economic and environmental considerations

If and how economic considerations, biological, ecological or scientific facts affected the location of the boundary, e.g. biological provinces, currents, environmental protection, economic power, economic need, unity of deposits, division of mineral and living resources, and joint development zones.

Geographic considerations

If and how geographic factors affected the location of the boundary, e.g. coastal configurations, concavities, convexities, general direction of the coasts, and location and direction of the land boundary.

Islands, rocks, reefs, and low-tide elevation considerations

If and how islands, rocks, reefs, and low-tide elevations affected the location of the boundary in regard to such matters as the entitlement of such features to maritime zones; the use of such features as basepoints for the delimitation of maritime areas attaching to larger land formations; and the significance of factors such as location (e.g. proximate, distant, mid-ocean, etc.), size, population, economic status, political status, or disputed status.

Baseline considerations

If and how the baseline from which the territorial sea is measured (such as bay and river closing lines, systems of straight lines, and lines delimiting historic or archipelagic waters) or other similar lines, affected the location of the boundary.

Geological and geomorphological considerations

If and how factors of geology and geomorphology, including the concept of 'natural prolongation', may have affected the location and extent of the boundary.

Method of delimitation considerations

What were the methods used to delimit the boundary line? If and how oppositeness, or adjacency and proportionality affected the delimitation.

Technical considerations

If and how technical issues of delimitation affected the location of the boundary, e.g. questions relating to map projections, simplifications, measurements of areas, distances, water datums, whether the boundary or part of it is ambulatory, and whether the coordinates of the boundary were computed or determined by geographical means.

This study endeavored to report on all of the considerations that contributed to the various boundary settlements and the common forms of behavior that could be identified. When hard information was unavailable, the authors of the boundary reports were invited to provide their own expert judgment on the considerations that entered into the settlements.

THE ANALYTICAL PAPERS

The boundary reports provided the foundation for the analytical papers that were commissioned by the project. These analytical papers are of two kinds. First, the Regional Experts were invited to prepare papers synthesizing the results of their study of the maritime boundaries in their regions and related areas. Regional papers covering each of the ten regions are included in this book. These papers identify certain facts that have influenced the delimitations in the individual regions. The regional analyses are important to a full understanding of the context and meaning of the information found in the individual boundary reports. They provide information on the historical and contemporary circumstances that may have influenced the delimitation. They also identify regional practices, if any, and the maritime boundaries remaining to be delimited.

Second, other experts, designated 'Subject Experts', were invited to analyze the boundary reports from a global perspective. Experts were assigned to examine each of the nine considerations identified above. Those experts were:

1 Political, strategic, and historical considerations – Bernard H. Oxman
2 Legal regime considerations – David Colson
3 Economic and environmental considerations – Barbara Kwiatkowska
4 Geographic considerations – Prosper Weil
5 Islands, rocks, reefs, and low-tide elevation considerations – Derek Bowett
6 Baseline considerations – Louis Sohn

The first paper that analyzes the individual considerations from a global perspective is Bernard Oxman's on the political, strategic, and historical considerations that may have influenced the delimitations. He found that there were few cases in which government representatives made it explicit that these considerations played any role. There is little doubt, however, that these considerations are in the background for all cases and often motivate settlement. It is the rare case, however, in which particular interests relating to maritime boundaries reach high levels of national or international importance. Thus, Oxman points out that states will seek negotiated settlements and are often willing to allow third-party settlements in order to avoid an escalation of these conflicts and the disruption of other bilateral or regional relationships.

While the boundaries studied involve a variety of legal regimes, it appears from the available information considered by David Colson that states rarely, if ever, are influenced by legal regime considerations. This is best illustrated in those cases where maritime boundary lines that were established for the purpose of one type of maritime jurisdiction have been applied to others without changing the location of the boundary lines. Neither have states faced particular difficulties negotiating their maritime boundaries when their maritime claims have differed in extent or scope.

Adjudications and arbitrations of maritime boundary disputes appear to place decreasing emphasis upon economic and environmental considerations. It is not clear that the maritime boundary agreements reflect the same trends. Barbara Kwiatkowska's study of these factors suggests that economic considerations may play important roles in such settlements. Economic interests in regard to current and historic fisheries, current and potential mineral resource development, and navigation certainly provide a leitmotif for the negotiation and conclusion of such agreements. In many situations in which such interests are present the boundary agreements or related agreements establish special regimes for particular economic activities. Evidence suggests that in a small minority of the agreements (particularly situations in which there have been ongoing mineral resource activities), these considerations have directly influenced the location of the boundary line. Even less of a role has been given to environmental and navigational considerations.

Prosper Weil draws a stronger distinction between the law binding on courts and arbitral tribunals charged with establishing a maritime boundary, and interstate negotiations to settle such boundaries. While third-party forums are required to apply the law, states are free to negotiate solutions on the basis of any considerations they wish. Nevertheless, those two modes of establishing maritime boundaries are closely linked in fact. The law applicable to third-party settlements will influence the negotiators, as will trends in boundary settlements influence the tribunals. In virtually all situations coastal geography is primary. Thus, the landmass, seabed, and human geography have limited relevance, if any, to such boundary delimitations. Even in the case of coastal geography its influence on the maritime boundary is not clear. It is difficult to determine how particular geographic facts influenced the location of a specific delimitation line. While the coastal geography is an objective fact, its importance to any particular delimitation is the result of subjective, not objective, reasoning. Finally, Weil reports that the available evidence suggests that there are no definite patterns of state practice; at most there are trends. Any analysis of the role of geography in these delimitations demonstrates that conflicting examples involving apparently similar geographic circumstances abound. As a consequence, the settlements are so varied that Weil dismisses the idea that the practice supports any normative rule.

Perhaps the most interesting boundary settlements involve islands, rocks, reefs, and low-tide elevations. As Derek Bowett reports, the practice varies substantially. Rocks, reefs, and low-tide elevations do not serve as a basis for entitlements to economic zone or continental shelf areas. They may, however, influence maritime boundary delimitations as appendages to larger territorial units. On the other hand, an island may serve as a basis for an entitlement to such zones. The treatment of islands in maritime boundary delimitations varies substantially, mostly due to the variety of circumstances in which they are found. Professor Bowett classifies the circumstances in which islands are relevant to these delimitations on the basis of their entitlements to maritime zones and the effects which islands have had on the locations of the boundary lines. While these effects vary substantially, some patterns which are worthy of consideration can be discerned from the practice. These patterns are influenced by such factors as the distances separating the features, comparisons of coastal lengths, questions of political status, population, and economic self-sufficiency.

Louis Sohn studied the relationship of that baseline to maritime boundary delimitations. He found that most boundary agreements do not specify what basepoints or portions of the baseline were used to generate the boundary line. Modern boundary agreements simply describe a line in the sea by lines connecting points that are described by latitude and longitude. Unless the line is clearly an equidistant line, the connection between the baseline and the delimitation is not easy to ascertain.

A boundary line that departs from the true equidistant line at its seaward limit may allocate to one coastal state an area that could not otherwise be placed within that state's jurisdiction because the area will be located beyond the distance from its coastline permitted by international law. Apparently, in this 'grey zone' the receiving state exercises the maritime jurisdiction which the granting state is permitted to hold under international law. As Sohn and Colson point out, few maritime boundary agreements make this clear and it is not settled that the transfer of authority is permitted by international law.

Keith Highet examines the use of geophysical factors in maritime boundary delimitations (e.g. geology and geomorphology). Only a small proportion of the agreed maritime boundary delimitations took account of such factors to any degree whatsoever, although in a very few unusual cases these factors were significant. No decision of the International Court or other international tribunal has established a maritime boundary on the basis of geophysical factors. The role of geophysical factors, Highet finds, reached its apex in the years following the *North Sea* judgment. The period subsequent to the Law of the Sea Conference and the *Libya/Malta* judgment has shown a decline in the attention given to geophysical factors in maritime boundary delimitations. Highet maintains that states negotiating maritime boundary agreements are substantially constrained by the law that is enunciated by the Court. Unlike other areas of international law, maritime boundary disputes often lead to international adjudication. States are reluctant, he maintains, to put forward negotiating positions that they could not credibly maintain before the Court.

Leonard Legault and Blair Hankey jointly explored the methods used to delimit maritime boundaries. While some maritime boundaries are delimited by unique methods, many make use of standard methods and their variants. Legault and Hankey describe these common methods beginning with equidistance, and proceeding to parallels and meridians, enclaving, perpendiculars, and parallel lines. Legault and Hankey explore whether any correlation exists between the methods used in the delimitations studied and the coastal relationships of oppositeness and adjacency. They find substantial use of the equidistance method

8

throughout. This is particularly striking in the case of opposite coasts and less so in adjacent situations. Even in the latter, up to one-third of the boundaries are based upon equidistance. But these statistics may be misleading if changes in the jurisprudence over time and the geographical circumstances presented are not factored in. These considerations suggest a decline in equidistance as the International Court of Justice relegated it to a subsidiary role and as the 1958 Law of the Sea Conventions became less salient. Furthermore, early delimitations may reflect easy cases that were more amenable to solutions based upon the equidistance method and its variants. The role of proportionality in maritime boundary negotiations was less easy to ferret out. Proportionality has played a significant qualitative role in the maritime boundary cases. Evidence exists to establish both a qualitative and quantitative use of proportionality in agreed delimitations, but due to methodological questions its role is more amenable to the former.

It should be apparent that the delimitation of maritime boundaries requires the consideration of many geographic, geodetic, hydrographic, and cartographic facts. Peter Beazley studied the treatment given to certain of these technical matters in the settlement of maritime boundaries. He focused on the sources of data used to locate the baselines and basepoints, the differences in vertical and horizontal datums used, the methods used to produce the boundaries, the nature of the lines used to join boundary turning points, the methods by which the terminal points of the boundaries were defined, and the accuracy attained by the delimitation methods. His paper argues that inattention to such matters may lead to serious problems, especially where precise locations are necessary for certain maritime activities. Many contemporary maritime activities require accurate maritime boundaries. Beazley suggests that serious difficulties could be avoided if the negotiators and international tribunals would resolve these technical matters as they develop the boundary line.

THE PROJECT'S RESULTS

It would appear from the global and regional papers and the individual boundary reports that no normative principle of international law has developed that would mandate the specific location of any maritime boundary line. The state practice varies substantially. Due to the unlimited geographic and other circumstances that influence the settlements, no binding rule that would be sufficiently determinative to enable one to predict the location of a maritime boundary with any degree of precision is likely to evolve in the near future.

There are, however, trends and practices that are substantial. Surprisingly, it appears from the practice that the equidistant line has played a major role in boundary delimitation agreements, regardless of whether they concern boundaries between opposite or adjacent states. In the vast preponderance of the boundary agreements studied, equidistance had some role in the development of the line and/or the location of the line that was established. Indications in some literature that suggest the demise of equidistance would appear to be incorrect.

Clearly, even today the focus upon coastal geography has had the effect of limiting the geographical range in which those boundaries may be found. States appear to be concentrating upon the division of water areas. Techniques of equidistance (with minor features disregarded or discounted) and other areal divisions suggest a narrow range of acceptable solutions to many maritime boundaries.

Geography, however, does not fully explain why many of the maritime boundary lines were drawn as they were. The available evidence suggests that geological, geomorphological, environmental, and strategic considerations have little role in these matters. Political, historical, and economic considerations may be more salient.

Often economic considerations are removed through arrangements that focus separately on those specific interests. But economic interests founded upon historic rights and usage remain as important elements in these negotiations. Continuing historical activities are, however, based upon facts that may be objectively determined. Those elements combined with geography could lead to a relatively clear norm. The studies suggest, however, that historical practices explain only a few of the results. Political considerations may explain some of the otherwise inexplicable results. This conclusion must be reached primarily on the basis of deduction since evidence in this regard was nearly impossible to garner. While politics appears to play a role, it should be clear, however, that its impact is limited. This is a result of the fact that the products of these negotiations are depicted as lines on maps which primarily represent the geographical context.

Geography and historical practices will increasingly direct their attention to the narrow range of alternative results that can be justified to the international community. While there is no lack of normative solutions, there is also no evidence of significant normative development. Perhaps this is due to the fact that negotiators have acted without knowledge of the practice throughout the world. If so, this study will begin to provide that information. More likely, states have not been satisfied that any particular relatively determinative norm would be satisfactory in all or the vast majority of cases. If such a rule could be designed, it might be embraced by the international community. At the present time, the legal focus is on whether the boundary has produced an equitable result. Although this is highly indeterminate, the study of the state practice provides some information about what the community regards as equitable.

THE FUTURE

Despite the normative and theoretical uncertainties present in this area, the evidence brought out in this study suggests that states seeking to delimit their maritime boundaries ought to consider certain facts and options as they develop their positions and resolve them through negotiation and third-party processes.

First, it is clear that primary attention will be placed upon the geography of the coastline.

Second, the equidistant line will be considered in most circumstances as a basis for analyzing the boundary situation. It may very well be used in some form or variant to generate the boundary itself.

Third, the delimitation of a definitive maritime boundary is not the only option available to states. While different boundaries for different regimes or uses are rare, creative settlements that take certain matters out of contention for boundary delimitation purposes are possible. Thus, joint development or management zones that cross boundaries, revenue sharing, and management cooperation are all possible options which, in the appropriate cases, can facilitate settlement or even make settlement of the maritime boundary irrelevant.

Fourth, even if a definitive boundary cannot be established, interim arrangements may be possible. States need to be careful to assure that such arrangements are, in fact and law, not unwittingly prejudicial to the ultimate settlement.

Fifth, a precise definition of the boundary line may be necessary at some point in the future. The boundary states would be well advised to memorialize their settlement in a technically precise form that would be unchallengeable in the future. The association of technical experts at the appropriate stages is strongly advised.

Sixth, the investment of substantial resources to study obscure geological and geomorphological facts may not be rewarding in either international litigation or negotiation. Knowledge of ongoing exploitation of living and nonliving resources by the boundary states in the boundary area, however, is likely to contribute to the development of a successful solution.

Seventh, the state representatives should bring to negotiations and litigation knowledge of the boundary states' international agreements that may bear on the location of the boundary line. This would include agreements relating to the terminus of the land boundary, and on the rights of the parties in the areas in question.

Eighth, maritime boundary delimitations cannot be divorced from the status of the general relations between the boundary states. Actively hostile relations will doom boundary settlement negotiations; less than solid friendly relations will make cooperative arrangements impossible to negotiate and implement.

Ninth, despite the relative indeterminacy of the maritime boundary law there are, in state practice and in judicial decisions, real limits to the geographical range in which a maritime boundary between two states will be located. These limits are primarily a function of the coastal geography, the size and location of islands, and the waters of the areas in question. What is ultimately considered to be fair or equitable will be largely dictated by a visual conception by the decision-makers of the maps and charts examined for this purpose. As a consequence, focus will be on the division of the water areas in question relative to the coastal states.

Finally, within the above considerations the law and practice permits states and tribunals a range of discretion that allows for the resolution of maritime boundaries in ways that no state need be characterized as a winner or a loser, unless a state were itself to stake out an unswervingly doctrinal position. Viewed in isolation, boundary making is a zero sum game. However, the options available to vary the line over extended distances and to resolve related issues on the basis of non-boundary solutions allows for the resolution of maritime boundaries to the optimal advantage of all the participants.

The results of this study are not the definitive work on maritime boundaries. Much is left to be done by states and scholars. While nearly 140 maritime boundaries have been established since 1940 many more, perhaps the most difficult of them, remain to be settled. It is hoped that the instant work will make a substantial contribution to an understanding of the field, lead to further scholarly work, and facilitate the just resolution of the remaining maritime boundaries.

THE ASIL MARITIME BOUNDARY PROJECT

NOTES

1 Funding was provided by grants from the Ford and Mellon Foundations with supplemental grants from the Amoco and the Mobil and the Exxon Foundations, for which the project participants are most grateful.

2 *International Maritime Boundaries*, edited by Charney and Alexander, 1992, Dordrecht: Martinus Nijhoff. Volume 1 – 1,228 pages, Volume 2 – 944 pages

3 See the list of boundaries studied in Appendix A.

4 With assistance from David Colson, Robert W. Smith, and Elizabeth Verville.

5 With assistance from J.R. Victor Prescott, Robert Smith, and David Anderson.

6 With assistance from David Colson.

7 With assistance from Giampiero Francalanci.

APPENDIX A

MARITIME BOUNDARIES STUDIED

I North America

1-1	Canada–Denmark (Greenland) (1973)
1-2	Canada–France (St Pierre & Miquelon) (1972)
1-3	Canada–United States (Gulf of Maine) (1984)
1-4	Cuba–United States (1977)
1-5	Mexico–United States (1976)
1-6	United States–Soviet Union (1990)

II Middle America and the Caribbean

2-1	Colombia–Costa Rica (1977)
2-2	Colombia–Dominican Republic (1978)
2-3	Colombia–Haiti (1978)
2-4	Colombia–Honduras (1986)
2-5	Colombia–Panama (1976)
2-6	Costa Rica–Panama (1980)
2-7	Cuba–Haiti (1977)
2-8	Cuba–Mexico (1976)
2-9	Dominican Republic–Venezuela (1979)
2-10	France (Martinique)–St Lucia (1981)
2-11	France (Guadeloupe & Martinique)–Venezuela (1980)
2-12	Netherlands (Netherlands Antilles)–Venezuela (1978)
2-13(1)	Trinidad and Tobago–Venezuela (Gulf of Paria) (1942)
2-13(2)	Trinidad and Tobago–Venezuela (1989)
2-13(3)	Trinidad and Tobago–Venezuela (1990)
2-14	United States–Venezuela (1978)
2-15	Dominica–France (1987)

APPENDIX A

III South America

3-1	Argentina–Chile (1984)
3-2	Argentina–Uruguay (1973)
3-3	Brazil–France (French Guiana) (1981)
3-4	Brazil–Uruguay (1972)
3-5	Chile–Peru (1952)
3-6	Colombia–Costa Rica (1984)
3-7	Colombia–Ecuador (1975)
3-8	Costa Rica–Ecuador (1985)
3-9	Ecuador–Peru (1952)

IV Africa

4-1	Cameroon–Nigeria (1975)
4-2	The Gambia–Senegal (1975)
4-3	Guinea–Guinea Bissau (1985)
4-4	Guinea Bissau–Senegal (1989)
4-5	Kenya–Tanzania (1976)
4-6	Mauritania–Morocco (1976)
4-7	Mozambique–Tanzania (1988)

V Central Pacific/East Asia

5-1	Australia–France (New Caledonia) (1982)
5-2	United Kingdom (Sarawak, North Borneo, Brunei) (1958)
5-3	Australia–Papua New Guinea (1978)
5-4	Australia–Solomon Islands (1988)
5-5	Cook Islands–United States (American Samoa) (1980)
5-6	Fiji–France (New Caledonia & Wallis and Futuna) (1983)
5-7	France (French Polynesia)–United Kingdom (Pitcairn, Henderson, Duce and Oeno Islands)(1983)
5-8	France (Wallis and Futuna)–Tonga (1980)
5-9(1)	Indonesia–Malaysia (Continental Shelf) (1969)
5-9(2)	Indonesia–Malaysia (Territorial Sea) (1970)
5-10	Indonesia–Papua New Guinea (1980)
5-11	Indonesia–Singapore (1973)
5-12	Japan–South Korea (1974)
5-13(1)	Malaysia–Thailand (Territorial Sea) (1979)
5-13(2)	Malaysia–Thailand (Gulf of Thailand Continental Shelf) (1979)
5-14	New Zealand (Tokelau)–United States (American Samoa) (1980)
5-15(1)	North Korea–Soviet Union (Territorial Sea) (1985)
5-15(2)	North Korea–Soviet Union (Exclusive Economic Zone and Continental Shelf) (1986)
5-16	Papua New Guinea–Solomon Islands (1989)
5-17	France–Solomon Islands (1990)
5-18	Cook Islands–France (1990)

15

VI Indian Ocean

6-1   Australia (Heard/McDonald Islands)–France (Kerguelen Island) (1982)
6-2(1) Australia–Indonesia (Seabed Boundaries) (1971)
6-2(2) Australia–Indonesia (Timor and Arafura Seas) (1972)
6-2(3) Australia–Indonesia (Papua New Guinea) (1973)
6-2(4) Australia–Indonesia (Fisheries) (1981)
6-2(5) Australia–Indonesia (Timor Gap) (1989)
6-3   Burma–India (1986)
6-4   Burma–Thailand (1980)
6-5   France (Réunion)–Mauritius (1980)
6-6(1) India–Indonesia (1974)
6-6(2) India–Indonesia (Andaman Sea & Indian Ocean) (1977)
6-7   India–Indonesia–Thailand (1978)
6-8   India–Maldives (1976)
6-9   India–Maldives–Sri Lanka (1976)
6-10(1) India–Sri Lanka (Historic Waters) (1974)
6-10(2) India–Sri Lanka (Gulf of Manaar and Bay of Bengal) (1976)
6-11   India–Thailand (1978)
6-12(1) Indonesia–Malaysia–Thailand (1971)
6-12(2) Malaysia–Thailand (Northern Straits of Malacca) Continental Shelf (1971)
6-13(1) Indonesia–Thailand (Malacca and Andaman Sea) (1971)
6-13(2) Indonesia–Thailand (Andaman Sea) (1975)

VII Persian Gulf

7-1   Abu Dhabi–Dubai (1968)
7-2   Bahrain–Iran (1971)
7-3   Bahrain–Saudi Arabia (1958)
7-4   Dubai–Sharjah (1981)
7-5   Iran–Oman (1974)
7-6   Iran–Qatar (1969)
7-7   Iran–Saudi Arabia (1968)
7-8   Iran–UAE (Dubai) (1974)
7-9   Qatar–UAE (Abu Dhabi) (1969)
7-10   Sharjah–Umm al Quaiwan (1964)

VIII Mediterranean/Black Sea

8-1   Cyprus–United Kingdom (Akrotiri, Dhekelia) (1960)
8-2   France–Italy (1986)
8-3   France–Monaco (1984)
8-4   Greece–Italy (1977)
8-5   Italy–Spain (1974)
8-6   Italy–Tunisia (1971)
8-7(1) Italy–Yugoslavia (Continental Shelf) (1968)

16

APPENDIX A

8-7(2)   Italy–Yugoslavia (Territorial Sea) (1975)
8-8   Libya–Malta (1986)
8-9   Libya–Tunisia (1988)
8-10(1) Turkey–Soviet Union (Territorial Sea) (1973)
8-10(2) Turkey–Soviet Union (Continental Shelf) (1978)
8-10(3) Turkey–Soviet Union (Exclusive Economic Zone) (1987)

IX Northern and Western Europe

9-1   Denmark (Faroe Islands)–Norway (1979)
9-2   France–Spain (1974)
9-3   France–United Kingdom (1978, 1982, and 1988)
9-4   Iceland–Norway (Jan Mayen) (1980 and 1980)
9-5   Ireland–United Kingdom (1988)
9-6   Norway–Soviet Union (1957)
9-7   Portugal–Spain (1976)
9-8   Denmark–Federal Republic of Germany (1969)
9-9   Denmark–Norway (1965)
9-10   Denmark–United Kingdom (1971)
9-11   Federal Republic of Germany–Netherlands (1964 and 1971)
9-12   Federal Republic of Germany–United Kingdom (1971)
9-13   Netherlands–United Kingdom (1965 and 1971)
9-14   Norway–Sweden (1968)
9-15   Norway–United Kingdom (1965 and 1978)
9-16   Belgium–France (1990)
9-17   Belgium–United Kingdom (1991)

X Baltic Sea

10-1   Denmark–Federal Republic of Germany (1965)
10-2   Denmark–Sweden (1984)
10-3   Finland–Sweden (1972)
10-4(1) Finland–Soviet Union (C.S. in Gulf of Finland) (1965)
10-4(2) Finland–Soviet Union (C.S. N.E. Baltic Sea) (1967)
10-4(3) Finland–Soviet Union (Fishing N.E. Baltic Sea) (1980)
10-4(4) Finland–Soviet Union (E.E.Z., Fishery Zone, C.S., N.E. Baltic Sea)
        (1985)
10-5   Federal Republic of Germany–German Democratic Republic (1974)
10-6(1) German Democratic Republic–Poland (1989)
10-6(2) Federal Republic of Germany–Poland (1990)
10-7   German Democratic Republic–Sweden (1978)
10-8   Poland–Soviet Union (1985)
10-9   Sweden–Soviet Union (1988)
10-10   Poland–Sweden (1989)
10-11   Denmark–German Democratic Republic (1988)
10-12   Poland–Sweden–USSR (1989)

17

2

STATE PRACTICE IN MARITIME DELIMITATION

Rodman R. Bundy[1]

INTRODUCTION

Since 1942, more than one hundred maritime delimitation agreements have been entered into between states. These agreements deal principally with the continental shelf, although some refer simply to the delimitation of 'maritime areas' between the states concerned. Now that the concept of the Exclusive Economic Zone has become established as a principle of customary international law, despite the fact that the United Nations Convention on the Law of the Sea has not yet come into effect, it can be expected that future delimitations will increasingly deal with the EEZ as well.

Collectively, these agreements fall under the rubric of 'state practice'. Before discussing a number of specific examples, it is perhaps appropriate to place state practice in its proper context by commenting briefly on its legal relevance.

THE LEGAL RELEVANCE OF STATE PRACTICE

References to state practice in the jurisprudence

The question here is two-fold: (i) whether state practice points to the existence of any particular method of delimitation that is legally obligatory or that has an *a priori* or privileged status over other methods; and (ii) if not, whether state practice is nonetheless still relevant in a legal context to maritime delimitation.

Here, the starting point must be what the International Court of Justice had to say in the first continental shelf delimitation case decided by it – the 1969 *North Sea Continental Shelf Cases* between Denmark, the Federal Republic of Germany, and The Netherlands. The Court made several important pronouncements of principle which have largely been followed by the subsequent jurisprudence.

First, the Court stated that 'there is no legal limit to the consider-ations which states may take account of' in delimiting their continental shelf.[2] In itself, this suggests that there is no single method of delimi-tation which is obligatory in all circumstances.

Next, the Court laid down two criteria that had to be satisfied for state practice to reflect a general principle of customary international law. First, state practice must amount to a 'settled practice' in the sense of being 'extensive' and 'virtually uniform' with regard to the method of delimitation employed. Second, evidence must exist that the states concerned considered that the use of a particular method of delimitation was rendered obligatory by the existence of a rule of law requiring it – in other words, that they were conforming to a legal obligation when they agreed to a particular boundary.[3]

With respect to the body of delimitation agreements taken as a whole, neither of these criteria are met. There is no 'settled practice' when it comes to delimitation methods that have been used by states. Similarly, there is no evidence that states have felt compelled to use any particular method because it was required by law.

This has been confirmed by the International Court of Justice in its 1985 judgment in the Libya–Malta case. There, both parties presented extensive arguments based on state practice. Malta argued in favor of an equidistance line and contended that state practice, even if it did not reflect a particular rule of customary international law, did provide significant evidence of normal standards of equity which pointed to the use of an equidistance or a median line boundary in cases of states with coasts lying opposite each other. Libya, on the other hand, argued that state practice did not point to any one method as obligatory and that, if anything, state practice showed a trend away from equidistance in recent years. The Court held that the practice cited fell short 'of proving the existence of a rule prescribing the use of equidistance, or indeed of any method, as obligatory'.[4]

Despite this limitation as to the legal relevance of state practice, the Court and other arbitral tribunals have had no hesitation in borrowing liberally from it to justify the use of particular methods of delimitation in different circumstances. Thus, state practice remains of importance for judicial decisions even though it does not point to any one method of delimitation as obligatory.

The UK–France Arbitration

The first such reference appeared in the 1978 arbitration award rendered in the case between the United Kingdom and France over the delimitation of their respective maritime areas in the English Channel and the Atlantic. A map depicting the Arbitration Court's final Award appears as Figure 2.1.

A particular problem was posed by the presence of the Channel Islands which lay significantly closer to the French coast of Normandy. The solution adopted by the Court was to enclave the islands partially so that they received a band of territorial sea in what was otherwise the French continental shelf area. The use of such enclaves is found fairly frequently in state practice as will be illustrated below. However, another problem was presented by the Scilly Islands, which lie some 21 miles off the British coast of Cornwall.

Had a strict equidistance line been used, the Scillies would have had a disproportionate effect on the seaward portion of the line. Accordingly, the Arbitration Court decided to give the Scillies a 'half-effect'. This was done by constructing two equidistance lines: one drawn from the coasts of the parties ignoring the Scillies; and a second using the Scillies as basepoints, thus giving them full effect. A line was then drawn half-way between these two lines, resulting in a 'half-effect'.

What is interesting is the way in which the Court of Arbitration justified this result. It stated:

A number of examples are to be found in State practice of delimi-

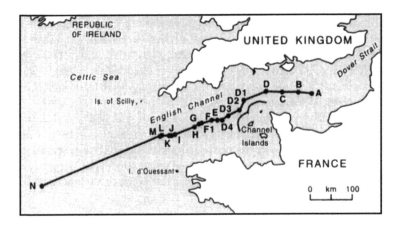

Figure 2.1 The France–United Kingdom maritime boundary

tations in which only partial effect has been given to offshore islands situated outside the territorial sea of the mainland ... in one instance, at least, the method employed was to give half, instead of full, effect to the offshore island in delimiting the equidistance line.[5]

The example that the Court referred to was, almost certainly, the delimitation between Iran and Saudi Arabia in which the important Iranian island of Kharg was accorded a half-effect in constructing what was otherwise an equidistance boundary. This agreement is discussed below.

The Libya–Tunisia continental shelf case

Following the United Kingdom–France Arbitration, the next decision of importance was the Libya–Tunisia case decided by the International Court of Justice in 1982. The resulting boundary is depicted on Figure 2.2.

The Court decided on a two-sector line. The first sector hinged on a number of historical and geographical facts, including the existence of an old Italian fishing boundary running roughly perpendicular to the coast and certain Libyan petroleum concessions.

In the second sector, the Court had to take account of the change of direction of the Tunisian coast which swings around the Gulf of Gabes from south to north before heading in a north-east direction towards Ras Kaboudia. Here, the Court was confronted with the question of what effect to give the Kerkennah Islands.

Ultimately, the Court decided to grant the Kerkennahs a 'half-effect'. This was done by drawing a line along the mainland coast of Tunisia from the most westerly point of the Gulf of Gabes (which produced an angle of 42°); then drawing a second line from the most westerly point along the seaward coast of the Kerkennahs (producing an angle of 62°); then splitting the difference (52°) and transposing this angle to the second sector of the delimitation line.

Once again, the reasoning of the Court reflects the importance of state practice. The Court said:

> The Court would recall ... that a number of examples are to be found in State practice of delimitations in which only partial effect has been given to islands situated close to the coast.... One possible technique for this purpose, in the context of a geometrical method of delimitation, is that of the 'half-effect' or 'half-angle'.[6]

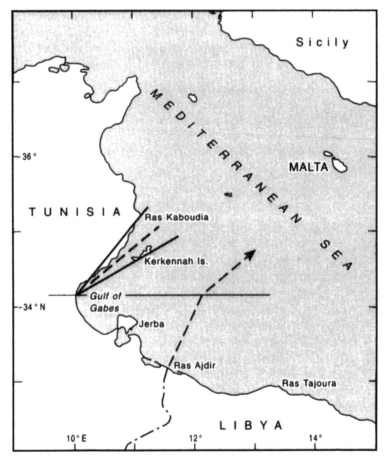

Figure 2.2 The Libya–Tunisia maritime boundary as decided by the
International Court of Justice (1982)

In the same judgment, the Court made another reference to state
practice – this time to the 1974 France–Spain delimitation in the Bay of
Biscay – as justification for holding that equidistance is not obligatory,
and that it can be combined with other methods of delimitation when
the circumstances so warrant in order to produce an equitable result.

The US–Canada single maritime boundary

The last case in this context is the Gulf of Maine case between the

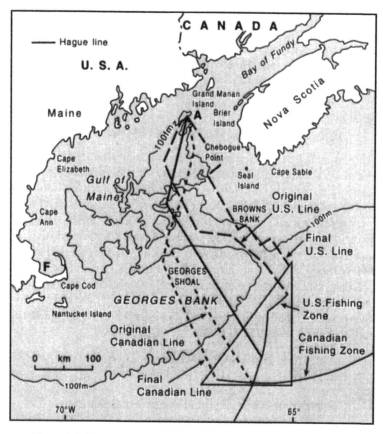

Figure 2.3 The Canada–United States boundary in the Gulf of Maine

United States and Canada decided by a Chamber of the World Court in 1984 (Figure 2.3).

Here again, it is the seaward segment of the delimitation line – between Nova Scotia, on the one hand, and the coasts of Maine, New Hampshire and Massachusetts, on the other – that merits attention. It can be seen that the coasts of the latter – that is, on the US side – are considerably longer than the corresponding coast of Canada along the Gulf of Maine.

The Court took this disparity of coastal length into account in arriving at the delimitation line. Without going into the details, the Court adjusted the seaward portion of the line closer to the Canadian coast by the same proportion (approximately 1:1.6) that the lengths of the respective coasts bore to each other. As justification for this adjustment, the Court referred to the Libya–Tunisia case and to examples of state practice, including the France–Spain Agreement, in which the difference in the length of the coasts of the states involved had influenced the resulting boundary. The Court also gave half-effect to Seal Island, an island lying off the south-east coast of Nova Scotia.

Conclusions as to the legal relevance of state practice

It can be seen that while state practice does not point to any single method of delimitation that is obligatory in all situations, it does provide important evidence of the way in which states have dealt with peculiar geographical, historical and other factors. In this sense, state practice retains legal and practical relevance. Even though it does not reflect a principle of customary international law, it shows that a number of delimitation methods, or combinations of methods, may be employed depending on the facts and circumstances of each case.

Having said that, let us turn to some specific examples of state practice. These are by no means exhaustive, but they form a representative sample of the kinds of methods that states have used to delimit their continental shelves in the past.

EXAMPLES OF STATE PRACTICE

Agreements employing equidistance or a median line

Despite the Court's pronouncement that the equidistance method has no privileged or *a priori* status, it is still apparent that states have frequently resorted to equidistance, particularly when there are no circumstances that would tend to have a disproportionate effect on an equidistance line or not be properly reflected by such a line.

One case where a median line is frequently utilised is between islands of roughly equivalent size. The following examples illustrate the use of equidistance in such situations. In each case, equidistance was specifically referred to as forming the basis of the delimitation in the agreements themselves:[7] (a) France (Réunion)–Mauritius (Figure 2.4), (b) Italy–Spain (Figure 2.5). Similarly, median lines often form the underlying basis of delimitation where mainland coasts of similar length are involved. Examples include (c) Finland–Sweden (Figure 2.6), (d) Canada–Denmark (Greenland) (Figure 2.7) – stated to be a median line with some adjustments – and (e) Cuba–Mexico (Figure 2.8). The agree-

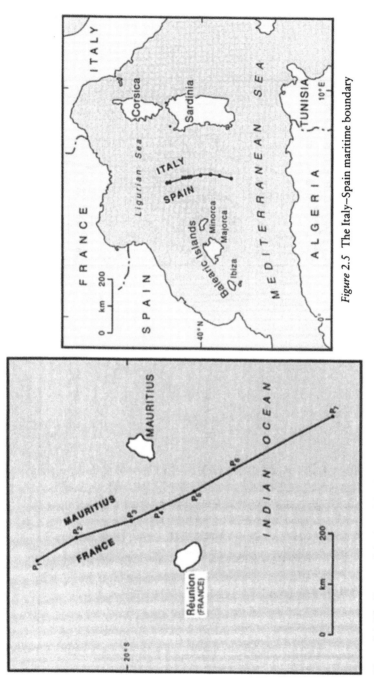

Figure 2.4 The France (Réunion)–Mauritius maritime boundary

Figure 2.5 The Italy–Spain maritime boundary

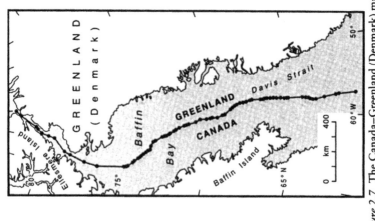

Figure 2.7 The Canada–Greenland (Denmark) maritime boundary

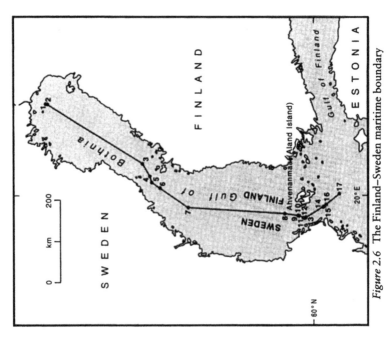

Figure 2.6 The Finland–Sweden maritime boundary

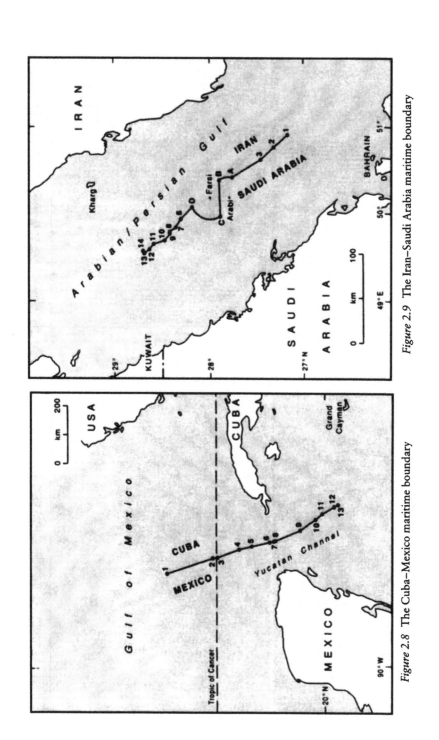

Figure 2.8 The Cuba–Mexico maritime boundary

Figure 2.9 The Iran–Saudi Arabia maritime boundary

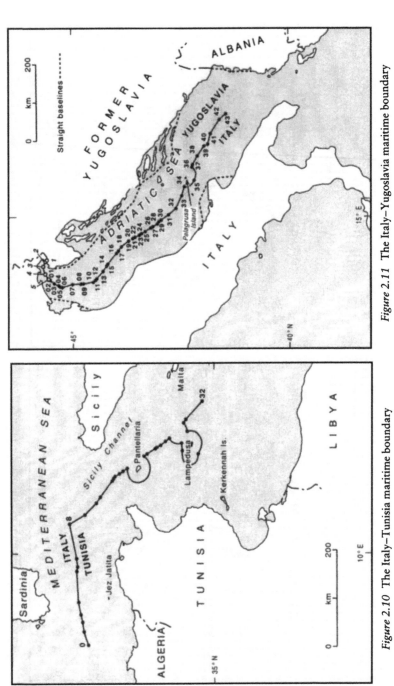

Figure 2.11 The Italy–Yugoslavia maritime boundary

Figure 2.10 The Italy–Tunisia maritime boundary

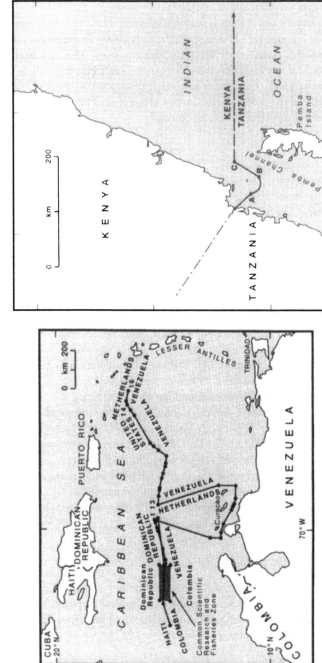

Figure 2.12 The Netherlands (Antilles)–Venezuela maritime boundary

Figure 2.13 The Kenya–Tanzania maritime boundary

ment states that the principle of equidistance was used. These examples are relatively straightforward. However, it is also apparent that equidistance is often more readily suited to delimitations between opposite coasts than between adjacent ones. This has been acknowledged by the International Court in its 1969 North Sea Judgment where the Court stated that as a general proposition there is less difficulty entailed in applying equidistance in the case of opposite coasts than in situations involving adjacent coasts.[8]

The treatment of islands

The situation becomes more complicated with the presence of islands. Because of its importance in the jurisprudence (having been referred to in several cases), the Saudi Arabia–Iran Agreement is worth mentioning first.

(a) Saudi Arabia–Iran (Figure 2.9). Half effect was accorded to Kharg Island according to the Geographer of the US State Department. In addition, twelve-mile partial enclaves were agreed for Farsi and Arabi Islands.

(b) Italy–Tunisia (Figure 2.10). Twelve- and thirteen-mile enclaves were agreed for the Italian islands; otherwise the delimitation was based on a median line.

(c) Italy–Yugoslavia (Figure 2.11). Twelve-mile enclaves were negotiated for Yugoslav islands of Pelagrusa and Kajola.

(d) Netherlands Antilles–Venezuela (Figure 2.12). Here, the Dutch Antilles did not receive full equidistance treatment, particularly in the northern sector.

(e) Kenya–Tanzania (Figure 2.13). This is an interesting example of state practice using a combination of methods, including a median line for the territorial sea in the Pemba Channel with an arc drawn around a lighthouse, and the use of a straight parallel of latitude in the seaward sector with Pemba Island having no effect on the seaward portion of the boundary.

Lines drawn perpendicular to the coast

It was seen that at least in the first sector of the Libya–Tunisia case, a line drawn 'normal' or perpendicular to the general direction of the coast was a relevant factor. A number of examples of state practice also illustrate the use of this method.

(a) Brazil–Uruguay (Figure 2.14). The agreement states that the boundary line is 'nearly perpendicular' to the general direction of the coast.

(b) Senegal–Guinea Bissau (Figure 2.15). A 240° azimuth or rhumb line has been employed roughly perpendicular to the coast.

Lines of longitude or latitude

This method is a close relative of the perpendicular method. Undoubtedly, one of its advantages lies in its simplicity. Examples include a number of delimitations along the west coast of South America as well as agreements in Africa and Central America: Chile–Peru (Figure 2.16), Peru–Ecuador (Figure 2.17), Colombia–Ecuador (Figure 2.18), Colombia–Panama (Figure 2.19) – this unusual boundary follows a step-like series of lines of longitude and latitude over part of its course – and The Gambia–Senegal (Figure 2.20).

The lengths of coasts involved

This factor was highly relevant in the Libya–Tunisia, Libya–Malta and US–Canada cases. Examples where the respective lengths of the coasts involved appear to have played an important role in the delimitation include France–Spain (Figure 2.21). This example was referred to in both the Libya–Tunisia and US–Canada cases. (The US State Department Geographer has referred to a coastal ratio of 1:1.54 in favor of France which may have played a role in the parties' agreement to push the seaward portion of the line closer to Spain.)

Geomorphology

In the 1969 *North Sea Continental Shelf Cases*, the Court observed that 'it can be useful to consider the geology of [the] shelf in order to find out whether the direction taken by certain configurational features should influence delimitation'.[9] Bearing in mind that at the time a state's legal title to the continental shelf rested in part on the notion of the 'natural prolongation' of the landmass into and under the sea, it was thought that geological or geomorphological features could play an important role in maritime delimitation, and parties to subsequent cases argued accordingly. Even in the 1982 Libya–Tunisia case, the Court left open the possibility that such features could be relevant to future delimitations.[10]

31

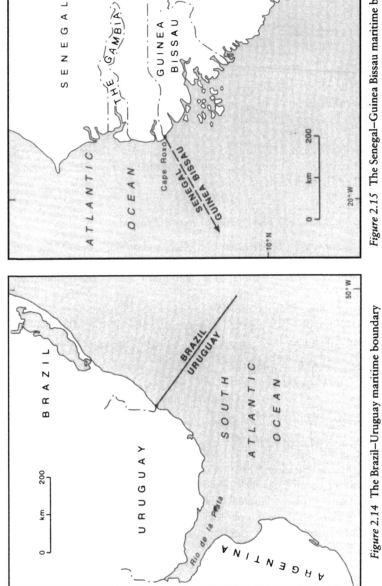

Figure 2.15 The Senegal–Guinea Bissau maritime boundary

Figure 2.14 The Brazil–Uruguay maritime boundary

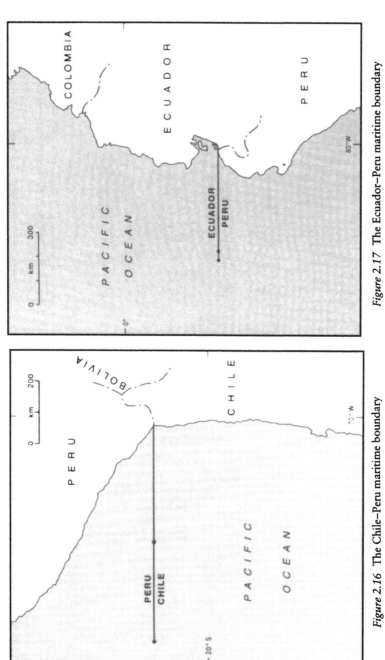

Figure 2.16 The Chile–Peru maritime boundary

Figure 2.17 The Ecuador–Peru maritime boundary

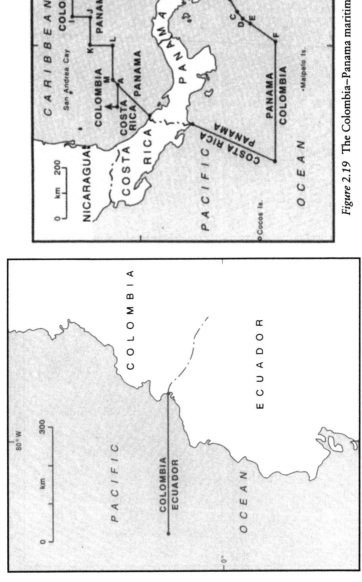

Figure 2.19 The Colombia–Panama maritime boundary

Figure 2.18 The Colombia–Ecuador maritime boundary

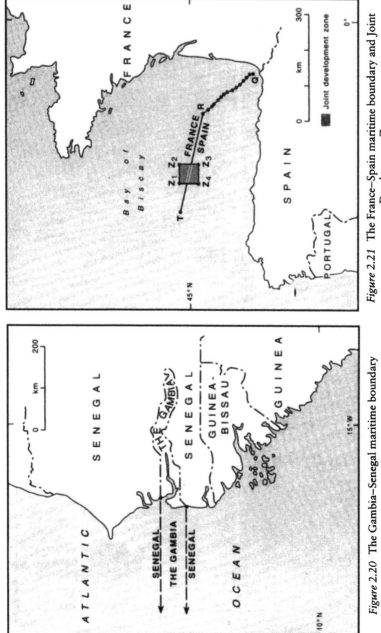

Figure 2.20 The Gambia–Senegal maritime boundary

Figure 2.21 The France–Spain maritime boundary and Joint Development Zone

In 1985, however, the Court took quite a different view. It noted that since its earlier decisions, the UN Convention on the Law of the Sea had been signed which provided for a distance criterion of 200 nautical miles to be used for determining a state's title to its continental shelf and exclusive economic zone. In these circumstances, the Court found that geological features had no role to play at least within 200 nautical miles of the baselines. Nonetheless, in at least two examples of state practice, geomorphological factors appear to have played a role:

(a) Australia–Indonesia (Figure 2.22). Equidistance was used in the east; but in the west there is evidence that both geomorphological factors – the Timor Trough – and the limits of Australian petroleum concessions were a factor.
(b) Saudi Arabia–Sudan (Figure 2.23). The parties agreed on a 'common zone' which was defined by depth criteria of 1000 metres.

Joint development zones

One potential problem encountered in maritime delimitation concerns the existence of mineral resources which straddle international boundaries. It is often the presence of such resources which provides the impetus for states to settle or adjudicate their disputes.

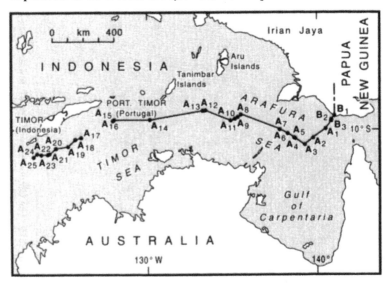

Figure 2.22 The Australia–Indonesia maritime boundary

36

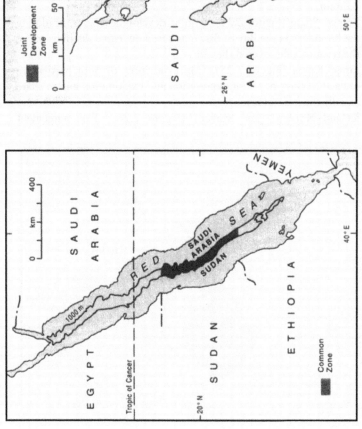

Figure 2.23 The Saudi Arabia–Sudan common zone

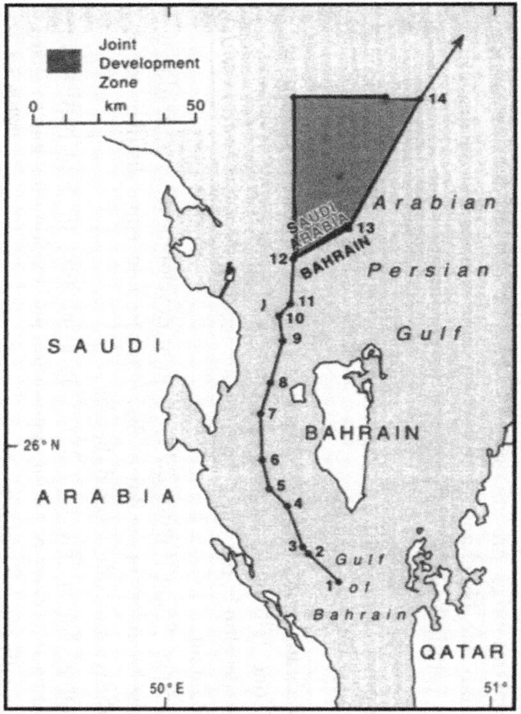

Figure 2.24 The Bahrain–Saudi Arabia maritime boundary and Joint Development Zone

How have states dealt with this problem? Two solutions find favor in state practice. In some instances, the agreements in question state that the parties will consult with each other regarding joint exploitation if mineral resources are discovered within a certain distance of the boundary line. No change in the delimitation is made, but full disclosure often permits the states involved to coordinate their development efforts.

In other cases, joint development zones have been created. Specific examples of these include:

(a) Bahrain–Saudi Arabia (Figure 2.24). The 'Joint Development Zone' falls under Saudi jurisdiction and Saudi Arabia is permitted to develop it, but must share the net income derived therefrom with Bahrain.

(b) France–Spain (Figure 2.21). A joint development zone straddling the boundary in the Bay of Biscay is based on 'equal distribution' of the resources discovered therein.

(c) Japan–Korea (Figure 2.25). Concessionaires of both parties are obliged to carry out joint exploration and exploitation and to share equally the natural resources.

(d) Iceland–Norway (Jan Mayen) (Figure 2.26). This agreement

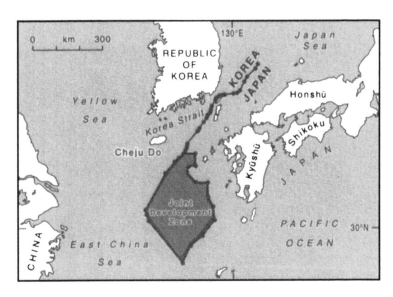

Figure 2.25 The Japan–Korea maritime boundary and Joint Development Zone

resulted from an international conciliation process. Iceland receives a 25% share of petroleum activities carried out in the Joint Economic Zone north of the delimitation line. Norway receives a 25% interest south of the delimitation line.

The presence of third states

Finally, delimitations in constricted areas where the presence of third states comes into play must be mentioned. As far as the jurisprudence is

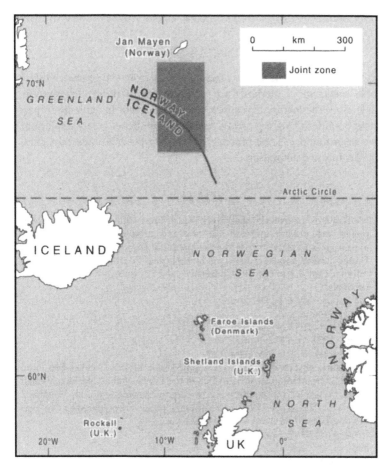

Figure 2.26 The Iceland–Norway (Jan Mayen) maritime boundary and Joint Zone

39

concerned, the Court has made it clear that the presence of third states is a relevant circumstance to be taken into account in effectuating delimitation. The Libya–Malta decision, which was severely limited on the east by Italian claims, bears this out.

In state practice, there are a number of regions which have been subject to more than one delimitation agreement. In some instances, such as in the North Sea or the Persian Gulf, there are no conflicts between the various agreements. In others, however, such as in the Caribbean, conflicts can arise. Thus, this is an additional factor that states must take into account in negotiating maritime boundary agreements.

CONCLUSIONS

It is apparent from the Court cases as well as from state practice that a whole spectrum of methods are available for maritime delimitation. While all of the factors that states consider relevant in agreeing to particular boundaries are not always known, a number of creative solutions have been found in state practice which may provide important guidance for future delimitations.

NOTES

1 Mr Bundy is a partner in the Paris office of Frere Cholmeley specialising in public and private international law and arbitration. Mr Bundy has appeared as Counsel before the International Court of Justice, the Iran–United States Claims Tribunal, the International Court of Arbitration of the International Chamber of Commerce and various *ad hoc* arbitration tribunals.
2 ICJ Reports 1969, p. 50, para. 93.
3 ibid., p. 44, para. 77.
4 ICJ Reports 1985, p. 38, para. 40.
5 Cmnd. 1978, p. 117, para. 251.
6 ICJ Reports 1982, p. 89, para. 129.
7 The figures reproduced here were adapted from Libya's Counter-Memorial to the International Court of Justice in the Libya–Malta Continental Shelf Case. (*Annex of Delimitation Agreements*, Counter-Memorial, Volume 2 Parts 1 and 2, October 1983.) The original maps were prepared by Mr Scott B. Edmonds of Maryland Cartographics.
8 ICJ Reports 1969, p. 88, para. 126.
9 ibid., p. 51, para. 95.
10 ICJ Reports 1982, p. 47, para. 44.

3

A HISTORY OF MARITIME BOUNDARIES ON NATIONAL OCEAN SERVICE NAUTICAL CHARTS

Charles E. Harrington

INTRODUCTION

The National Ocean Service (NOS) is the lead agency for the portrayal of maritime limits of the United States of America because of its responsibility to chart the nation's coastal waters. The 1958 Geneva Convention on the Territorial Sea and the Contiguous Zone states: 'the normal baseline for measuring the breadth of the territorial sea is the low water line along the coast as marked on large-scale charts officially recognized by the coastal state'. In 1976, NOS was requested to show various maritime limits on its regular issue of nautical charts. This chapter discusses the history of maritime boundaries on NOS charts, methods used in constructing the various maritime limits, the definition of the limits, the push for lateral seaward boundaries, and the technical aspects of maritime limits.

In the past two decades, there has been an increasing interest in coastal zone management, offshore oil and gas exploration, fisheries and maritime environmental conservation, and development of natural resources. These elements have placed pressure on federal and state governments to define their maritime limits. Many coastal states have not defined their maritime limits with their adjacent state or states. As of 1980, ten out of eighteen maritime coastal state boundaries, or portions thereof, remained unresolved. On the federal level, 'The United States may have to negotiate approximately 30 maritime boundaries that will account for approximately 10 percent of the total maritime boundaries of the world' (Smith 1981:397).

As a result of the increased interest in the offshore areas of the United States, NOS has had to take a leading role in portraying maritime limits on its nautical charts.

CHARLES E. HARRINGTON

DEFINITIONS

(i) *Territorial sea* – a band or belt of sea adjacent to a state's coast, beyond its land territory and its internal waters, over which it has complete sovereignty. This sovereignty also extends to the air space over the territorial sea as well as to its bed and subsoil. This sovereignty is exercised subject to the provisions of the Conventions of the Law of the Sea adopted by the United Nations' Conference at Geneva, 1958, and to other rules of international law. Prior to December 27, 1989, the territorial sea of the United States was 3 nm. The limit of the territorial sea is now 12 nm.[1]

(ii) *Contiguous zone* – a zone of the high seas contiguous to the territorial sea of a state. The coastal state may exercise control to (a) prevent infringement of its customs, fiscal, immigration, or sanitary regulations within its territory or territorial sea; and/or (b) punish infringement of the above regulations committed within its territory or territorial sea. The contiguous zone limit is 12 nm from the baseline from which the breadth of the territorial sea is measured.

(iii) *Natural resources limit* – an area extending seaward from the US coastline in which Puerto Rico, Texas, and Florida (Gulf of Mexico side only) are entitled to all lands, minerals, and other natural resources. The limit of this area is 3 marine leagues (9 nm) from the baseline from which the breadth of the territorial sea is measured.

(iv) *Exclusive economic zone* – a zone of the high seas contiguous to the high seas over which a coastal state may assert certain sovereign rights over natural resources. This limit is 200 nm from the baseline from which the breadth of the territorial sea is measured.

HISTORY

NOS began showing maritime limits on nautical charts in 1976. At the request of the Ad Hoc Committee on the Delimitation of the United States Coastline (commonly referred to as the Baseline Committee[2] (BC)) and the US Coast Guard, NOS began showing the territorial sea and contiguous zone limits, 3 and 12 nautical miles, respectively, on its charts. From 1970 to 1976, NOS was portraying the 3- and 12-nm limits on black-and-white 50-per cent reductions of its nautical charts. The changeover to portray the limits on the regular issue charts was not immediate. The limits were added to nautical charts as they came up for printing on their normal printing schedules. The printing schedules vary from six months for charts covering the busier and ever-changing harbors to twelve years for those charts on the North Slope of Alaska. There remains one chart, which covers a portion of the North Slope of Alaska, which still requires 3- and 12-nm limits after sixteen years.

The 200-nm limit began appearing on NOS nautical charts around 1977 after the enactment of the Fishery Conservation and Management Act (FCMA) of 1976, which was effective March 1, 1977. The limit was originally labelled on the charts as the Fishery Conservation Zone (FCZ). Shortly after President Reagan signed the Exclusive Economic Zone Proclamation of March 10, 1983, the 200-nm limit label was changed to the Fishery Conservation Zone–Exclusive Economic Zone (EEZ). This was done at the request of the National Marine Fisheries Service (NMFS) because so many of the current regulations were written under the title of the FCMA. Since then the regulations have been revised, and the FCZ label is being removed (just the EEZ label will be shown).

In 1982, NMFS requested the National Ocean Service to depict a 3-marine league (9 nm) natural resources limit for domestic fishery enforcement purposes on its nautical charts. NOS brought the request before the BC for discussion. Although the issue is completely under the jurisdiction of NOS, the BC does take an interest in all maritime limits depicted on United States charts. (NOS and Department of Commerce (DOC) appreciate the expertise of the BC and have found it beneficial in resolving boundary portrayal problems.)

METHODS

The 3- and 12-nm limits were originally placed on NOS charts by the Geographer's Office, Department of State (DOS), in the late 1960s. The arcs were pencilled in manually by the Geographer and presented to the BC for approval. They were then forwarded to NOS where the lines were transferred in ink to another copy of the chart. Both copies were returned to the Geographer and the BC for final approval. The pencilled copies were retained by the Geographer, and the inked versions were returned to NOS. NOS was responsible for printing and distributing the 50-per cent black-and-white reductions. At this stage, a caution note was added stating that the chart was not to be used for navigation, along with an explanation regarding the preparation and function of the territorial sea and contiguous zone limits. This process involved 160 of the 975 nautical charts issued by NOS. The limits are shown on only one chart scale covering an area. On the east and gulf coasts, the limits

CHARLES E. HARRINGTON

are shown on a continuous series of charts at 1:80,000 scale. For the west coast, Alaska, Hawaii, and the US territories and possessions, the chart coverage varies from 1:50,000 to 1:1,023,188 scale. The 200-nm limit was compiled mathematically utilising NOS computer equipment. Geodetic software was modified to allow us to compute geodetic points (at a specified interval, e.g. 30 minutes of one degree of arc) on an arc 200 nm from a salient baseline point. At various chart scales, connecting these points provided a smooth arc with a radius of 200 nm. Required input for each arc includes a baseline point geographic coordinate, a beginning and ending azimuth, a specified interval for points along the arc, and the distance from the baseline point. The output for the computation of each arc was a punch card with latitude and longitude, azimuth, and an identification designator. The cards and magnetic tape were submitted to the NOS computer facility with a program to convert the points into a plotter format to produce an overlay for each nautical chart. The lateral boundaries between the US and adjacent states were provided by DOS as published in the Federal Register. A total of 56 NOS nautical charts portray portions of the EEZ limit.

Under the Submerged Lands Act, the natural resources limit was granted to Florida (Submerged Lands No. 5) (Gulf of Mexico coast only), Texas (Submerged Lands No. 2), and Puerto Rico (Public Law). These limits are placed on the charts manually in the same manner and follow the same approval procedures as the territorial sea and contiguous zone limits. A total of 29 nautical charts covering these three areas show the 3-league limit (Puerto Rico 4, Texas 9, and Florida 16).

CURRENT PROCEDURES

The 1958 Geneva Convention on the Territorial Sea and Contiguous Zone states that 'the normal baseline for measuring the breadth of the territorial sea is the low water line along the coast as marked on large-scale charts officially recognized by the coastal state'. The purpose of this chapter is not to expound on the methods used to develop the low water line, but to point out that it is a line in constant change and, therefore, the 3- and 12-nm limits must be changed accordingly.

The process of change begins at the chart compilation level. The cartographers apply various sources to the chart drawing, a mylar copy of the chart on which the cartographers apply corrections, and the drawing is then sent to the Reproduction Branch for negative engraving. These sources may include our own hydrographic surveys, shoreline

44

manuscripts (which are compiled from tide-coordinated aerial photographs), US Army Corps of Engineers' channel surveys, and other federal, state, or private sources. After the cartographer corrects the low water line or the shoreline from which the 3- and 12-nm limits are constructed, a copy of the drawing is made and sent to me. The writer, in turn, manually constructs 3- and 12-nm arcs or bay closing lines using the corrected low water line. The corrected drawing and a paper copy of a current edition of the nautical chart are sent to DOS where Dr Robert Smith, Special Assistant for Ocean Affairs and Policy Planning, reviews my work. He then writes a memorandum describing the changes and submits it to the Chairman of the BC. The Chairman notifies the BC members about charts that need to be reviewed, along with any other related issues that need to be discussed.

TECHNICAL ASPECTS

The 3-, 9-, and 12-nm limits are placed on the nautical charts manually. The procedures used to establish the limits on NOS nautical charts have already been described. Because the low water line is constantly changing, the maritime limits will also change.

As a chart nears the reprinting cycle, it is examined for changes in the low water line or any shoreline change that may cause the limits to move. A copy is sent to my office for examination, where I manually pencil the corrections on the updated chart drawing. Because the charts are legal documents, the accuracy of the limits is an essential factor. In September 1983, the BC decided as a 'rule of thumb' that a new edition would be issued if the change was at least one-half the width of the line on the chart. The line weight on the chart is 0.5 mm (0.020 inches). At a scale of 1:80,000, the width of the line is equal to 40 m on the surface. If the line moves as little as 20 m, a new line is constructed. No consideration was given to the various chart scales used. For example, at 1:500,000 scale, the line weight would equal 250 m at half a line weight which measures only 0.25 mm on the chart. The majority of charts portraying the 3-, 9-, and 12-nm limits are at scales ranging from 1:80,000 to 1:200,000.

Within NOS, discussions have been held regarding the placement of 3-, 9-, and 12-nautical maritime limits in the automated database for nautical charts. The method for accomplishing this has not been resolved. Basically, two methods have been considered: (i) digitising the lines off the chart drawings and (ii) incorporating software into the system to compute the limits from salient points on the low water line.

Many questions remain to be answered before going too far. How will the limits be maintained? How will the bay closing lines be drawn? Who will maintain the limits? Can the BC be assured of the accuracy of the limits? Can NOS continue to adhere to the principles and policies used by the BC? The BC has already accepted the computer in the determination and application of the 200-nm limit. This should open the door for NOS to incorporate the remaining maritime limits in digital form and instill some confidence in the BC that it may be an acceptable process.

LATERAL SEAWARD BOUNDARIES

'As a consequence of 200-nautical mile maritime claims, every coastal country in the world will eventually have to negotiate at least one maritime boundary with at least one neighboring country' (Smith 1981:397). Thirty maritime boundaries may have to be negotiated by the United States; 10 off one or more of the 50 states, and 20 located off the coast of the American territories.

> Nine of the 10 boundaries off the 50 states will involve five different foreign neighbors; Canada, Cuba, Mexico, Russia, and The Bahamas. At the beginning of 1981, the United States had reached agreements or understandings for some kind of maritime boundary with Canada, Cuba, Mexico, and Russia. No boundary talks were held with The Bahamas.
>
> (Smith 1981:397)

NOS has provided consultation, computation, and charts to DOS in most, if not all, of these agreements. In the Gulf of Maine case involving Canada, NOS detailed one of its cartographers to DOS for approximately six months. The cartographer also went to Europe with a team of US attorneys handling the case before the International Court of Justice.

There have been numerous cases involving federal vs. states' rights in the past decade: (i) Massachusetts, regarding the closing lines of Nantucket Sound; (ii) Alaska, regarding the closing lines and islands of the North Slope; (iii) Mississippi, Alabama, and Louisiana regarding the closing lines of Mississippi Sound; (iv) Kotzebue Sound, Alaska, regarding a low water feature that affected a 24-nautical mile bay closing line; and (v) low water features off the coast of South Carolina that put a sunken wreck either just inside or just outside the 3-nautical mile limit. Several coastal states are in negotiations over their lateral seaward boundaries. Some, such as Maine–New Hampshire and

Georgia–Florida, have recently settled on positions for their lateral seaward boundaries. Others, such as California–Oregon and New York–Rhode Island, have been settled for a number of years. Georgia and South Carolina are settling on the boundary in the Savannah River with assistance from NOS.

Generally, NOS does not show the lateral boundaries between the states on its nautical charts. The states of Maine and New Hampshire did request the addition of the adjudicated limit from Portsmouth Harbor to the Isle of Shoals. NOS complied, and the limits are shown on three nautical charts. In my opinion, it is not a major problem to show this type of boundary. However, there will not be a major push to show them. In almost 180 years, there have been only two other cases where NOS was asked to show a maritime boundary between two states. The states of Maryland and Virginia had some differences of opinion in the Chesapeake Bay over oyster beds, and California and Nevada had some difficulty as to where the state boundary in Lake Tahoe was located. NOS now shows both boundaries on the charts covering those two areas. State boundaries and the US–Canada international boundary are shown on NOS charts of the Great Lakes.

During the period of 1908 to 1930 our nautical charts portrayed the 'A–B' line in the Dixon Entrance (a body of water separating the United States and Canada on the west coast of North America). The Canadians claim that the Alaska Boundary Tribunal Award of 1903 gave them a maritime limit, whereas the United States contends the A–B line represents a hypothetical line defining the territories, not the maritime areas. The placement of the A–B line would prevent the US islands of Dall, Prince of Wales, and other territory north of the line from enjoying a territorial sea or contiguous zone south of the A–B line. In 1929, DOS requested our Agency to remove the line from the nautical charts. Canadian charts still show the A–B line, and US charts show an equidistant line in Dixon Entrance, which leaves another boundary to be resolved.

SUMMARY

Fifteen years ago, NOS charts rarely depicted a maritime boundary of any type. With today's ever-expanding culture, the federal government, coastal states, ocean research groups, and pivate and public organisations want to know more precisely the limits of areas of potential economic value. The development of more sophisticated surveying and positioning equipment, along with increased controls on national resources and the environment, have brought on the need to know the boundary of an area being regulated, as well as who has jurisdiction over the area. NOS has the mandate to produce and maintain accurate, up-to-date oceanographic products, continue collecting data, continue building digital databases, and to do whatever it can to serve those involved in the oceanographic and coastal environments.

Currently there are at least two pieces of legislation in our Congress to give the states more jurisdiction as well as additional revenue. One is a follow-up to the presidential proclamation giving the United States a 12-nautical mile territorial sea and, in turn, extending the states' jurisdiction to 12 nm as well. The other is the Ocean and Coastal Resources Enhancement Act, which would establish an ocean and coastal resources enhancement fund and a coastal zone impact assistance fund. Both would be administered by the Secretary of Commerce. Basically, this legislation would give the coastal state and the local communities a portion of any revenue generated within 200 nm of the US coast. One of the requirements of this bill would be for the states to have an extended seaward lateral boundary out to 200 nm. Many of the states have had difficulty working out a seaward 3-nm lateral limit, so one can imagine what problems a 200-nm line might create.

NOTES

1 On December 27, 1989, past President Ronald Reagan signed a presidential proclamation extending the US territorial sea from 3 to 12 nm. The 3-nautical mile limit remains on the nautical charts because the proclamation did not alter existing federal or state law. The label on the limit was changed to '3-Nautical Mile Limit' to reflect state jurisdiction.

2 The BC was established in 1979 as a spin-off from the Law of the Sea Task Force to review questions relating to boundary demarcation of the United States, and to identify the baseline from which the offshore boundaries of the United States can be delineated. The BC consists of representatives from DOC, Department of the Interior, Department of Justice, Department of State (DOS), and the Department of Transportation. The BC meets on an as-needed basis. Since its conception, the BC has met an average of six times a year.

REFERENCES

Smith, Robert W. (1981), 'The maritime boundaries of the United States', *Geographic Review*, vol. 4, p. 397.

Submerged Lands Act, No. 5, 43 USCA No. 1313.

Submerged Lands Act, No. 2(c), 67 stat. 29, 43 USC No. 1301(c).

Public Law 96-205, Title VI, No. 606(a). 94 stat. 91, as amended March 12, 1980.

4

DELMAR
A computer program library for the
DELimitation of MARitime boundaries
Galo Carrera

INTRODUCTION

Since 1958, the time of the Geneva Conference on the Law of the Sea, there has been a need for a tool to perform rapidly and accurately many of the geodetic tasks associated with the modern delimitation of international maritime boundaries and offshore limits. This need has become evident by the increasing interest shown among developing nations to manage their natural resources in view of the 1982 United Nations Convention of the Law of the Sea.

The remarkable improvements in accuracy made recently in extraterrestrial geodetic positioning techniques also have to be taken into consideration in the making of modern boundaries. The Global Positioning System (GPS) and the Global Navigation Satellite System (GLONASS) either are capable now, or are expected soon to perform kinematic positioning at sea anywhere on the surface of the earth with an accuracy better than 100 metres. The ability to locate a vessel at sea with such an accuracy at ever-decreasing costs, demands refinements in the geodetic methodology used to determine maritime boundaries. DELMAR has been created in order to meet these higher standards of accuracy.

The ultimate purpose of DELMAR is to serve interdisciplinary teams under the leadership of international lawyers as a source of technical information during the analysis, negotiation, and verification of maritime boundary agreements. These teams might be configured to include experts, for example, in fields such as international law, national security, international relations, computer science, geodesy and hydrography, natural resources, and the environment.

Version 1.0 of this computer program library is the first step towards what hopefully will become a useful geodetic tool for the equitable delimitation of international maritime boundaries. Comments and suggestions from users will play a crucial role towards this end.

WHAT IS DELMAR?

DELMAR is a library of computer programs developed in support of the delimitation of international maritime boundaries on the surface of a geodetic reference ellipsoid. Its structure allows the user to select from a number of options which provide valuable technical information relevant to the preparation, actual negotiation, and verification prior to ratification of international boundary agreements. DELMAR may also prove valuable as an academic research tool for investigating the technical nature of existing agreements.

WHAT ARE THE MINIMUM SYSTEM REQUIREMENTS FOR OPERATING DELMAR?

DELMAR is written in the ANSI Fortran-77 and ANSI-proposed C languages. The following are its minimum hardware and software requirements:

- an IBM or IBM-compatible XT or AT Personal Computer;
- the Disk Operating System DOS version 3.1 or higher;
- a minimum of 512 K of RAM;
- a standard (360 K) or high-capacity (1.2 M) disk drive;
- a hard disk drive;
- a Numeric Data Processor (NDP) Intel 80X87, and
- a Monochrome Display Adapter (MDA) and a monochrome monitor.

A special effort has been made to avoid the need for a graphics display adaptor e.g. Hercules, CGA, EGA or VGA to run the numeric applications of DELMAR. However, one of these adaptors is required if the graphics capabilities of this library are to be exploited fully.

WHAT COMPUTING EXPERIENCE IS NEEDED TO USE DELMAR?

DELMAR is designed for use by those who are familiar with a Personal Computer (PC) and with Microsoft's Disk Operating System (DOS). It is assumed that the user has at least a basic familiarity with DOS.

Information about DOS can be found in standard DOS reference manuals and in Norton (1985) and Duncan (1986), among others.

WHAT LEGAL EXPERIENCE IS REQUIRED TO USE DELMAR?

DELMAR is likely to be best put to use by those who are familiar with the documents and concepts of the 1958 Geneva and 1982 United Nations (UNCLOS III) Conventions of the Law of the Sea. It is assumed that the user, or an individual within an interdisciplinary user group, understands the definitions of terms such as territorial sea, and contiguous and exclusive economic zones, as used in UNCLOS III. Additional information can be obtained, for example, from Shalowitz (1962), Jagota (1985), Prescott (1985, 1987), Blake (1987) and Johnston (1987), and by consulting articles found in numerous legal periodicals such as the *International Journal of Estuarine and Coastal Law* and the *Ocean Development and International Law Journal.*

WHAT GEODETIC BACKGROUND IS RECOMMENDED FOR THOSE INTENDING TO WORK WITH DELMAR?

DELMAR is a suite of computer programs that make use of the mathematical apparatus available to geodesists for the tasks of solving positioning problems on the surface of a reference ellipsoid. Concepts and terms such as geodetic ellipsoid and its coordinates, datum transformation parameters, and geodesics must be clearly understood. Additional conceptual geodetic information can be found, for example, in Vanicek and Krakiwsky (1986).

IS KNOWLEDGE OF BOUNDARY AGREEMENTS EXPECTED FROM THOSE USING DELMAR?

Previous exposure to the legal and technical nature of existing boundary delimitation agreements is an essential asset to the application of this computer library. From the legal point of view, this body of information forms what is referred to as the source of customary law. From the technical point of view, these agreements provide a framework in which the definition and interpretation of many technical terms and techniques can be found. In addition to these primary sources, the reader can find a comprehensive compilation of international boundary agreements and offshore claims in the publication *Limits in the Seas*, issued from time to time since 1970 by the State Department of the US Government. Valuable technical information can also be found, for example, in Beazley (1978) and Prescott (1985, 1987), as well as in the hydrographic literature, such as the *International Hydrographic Review* and the *Hydrographic Journal.*

WHAT IS THE STRUCTURE OF THE DELMAR PROGRAM LIBRARY?

DELMAR's key features are an on-line tutorial, and a collection of modules for the following applications: computation of maritime areas; determination of offshore limits; delimitation of equidistant boundaries; delimitation of weighted boundaries that include 'partial effects'; solution of peripheral tasks relevant to a boundary delimitation; a full-screen text editor for editing and viewing data files; graphics programs for the display of geographic data (Figure 4.1).

WHAT IS THE ON-LINE TUTORIAL?

The on-line tutorial is a computer program designed as a didactic tool. It contains information about the structure and use of software. The tutorial will prove most valuable to experienced users who want to obtain direct experience with the program as quickly as possible. Others may feel more comfortable once they have been exposed to a more detailed account of the program library through its printed manual. In any case, both the manual and the on-line tutorial complement one another.

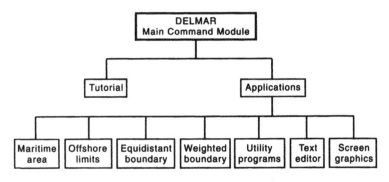

Figure 4.1 DELMAR program structure

52

WHAT OPTIONS ARE CONTAINED IN THE APPLICATIONS MODULE?

The Applications module contains all the computational sub-modules of the computer library. Each of these sub-modules is described in the following paragraphs. A technical description of the algorithms employed in all of the modules is given in the printed manual.

Maritime areas

The Maritime Area module is formed by two programs: *Block* and *Polygon*. Their purpose is to provide the area of a body of water on the surface of a geodetic ellipsoid, expressed in both kilometres squared and nautical miles squared.

Block determines the area contained in a section of a geodetic ellipsoid bounded by two parallels and two meridians. This program may prove useful to determine the maritime area of a 'rectangular' section such as those frequently employed in oil licences. Polygon is a more complex program. It determines the area of a closed polygon of any shape. The number of input vertices is limited to 640K by the memory of the computer itself. Furthermore, the sides of the polygon are automatically formed by geodesic lines. This program should produce the most accurate value of the area of a body of water along the coast when the coastline is to be followed very closely. Another potential application of this program is to determine the land and maritime areas inside a group of baselines used by a state claiming archipelagic status.

Offshore limits

The Offshore Limits module is formed by three programs: *TS-12*, *EEZ-200*, and *User-Limit*. TS-12 and EEZ-200 determine the position of the offshore limits located at distances of 12 and 200 nm respectively from points located on the low water line. These programs may find use when, in accordance with Articles 3 and 57 of UNCLOS III, a user wants to determine the offshore limits of the Territorial Sea (TS) and the Exclusive Economic Zone (EEZ). The output information of these programs is used by the program *Sweep* in the Utility module to describe the arcs with a user-prescribed number of points.

User-Limit allows us to determine the position of the offshore limits located at a user-prescribed distance from points located either on the low water line, the foot of the slope, or the 2,500-metre isobath. This program lends itself to a number of applications in accordance with Articles 3, 33, 57, and 76 of UNCLOS III:

- a 12 nm or a smaller user-prescribed distance for a territorial sea limit;
- a 24 nm or a smaller user-prescribed distance for a contiguous zone limit;
- a 200 nm or a smaller user-prescribed distance for an exclusive economic zone limit;
- a 60 nm or a smaller user-prescribed distance for a continental shelf limit from the foot of the slope;
- a 100 nm or a smaller user-prescribed distance for a continental shelf limit from the 2,500 m isobath; and
- a 350 nm or a smaller user-precribed distance for a continental shelf limit.

User-Limit produces exactly the same results as those given by TS-12 and EEZ-200 when the prescribed distances of TS-12 and EEZ-200 are 12 and 200 nm, respectively.

Equidistant boundaries

The Equidistant Boundary module is formed by the programs *Equid-23*, *Triplet* and *Pair*. Equid-23 is a revised and improved version of the program *Boundary*, which was developed by Carrera (1987) for a Digital Equipment VMS Vax computer environment. This program determines the equidistant boundary from two opposite or adjacent coastal states on the basis of the principle of equidistance as stated in the guidelines proposed by the International Law Commission (ILC) since 1953 (United Nations, 1953) and implied in Articles 12 and 6 of the 1958 Geneva Conventions of the Territorial Sea and the Contiguous Zone and the Continental Shelf respectively, and Article 15 of the 1982 Convention of the Law of the Sea. This principle states that equidistance must be measured from the nearest points from which the breadth of the territorial sea is measured. An alternative but equally important application of this program is the determination of an equidistant boundary as a reference for further negotiation, not as a final goal in an agreement.

Triplet and Pair are two components of Equid-23 which determine equidistant boundary turning points from pairs and triplets of low water line, respectively. They are included in order to allow the user to perform these operations separately. There are various examples of actual negotiations where this option would have been useful.

Boundary delimitations between adjacent coastal states produce occasional cases in which a rigorous implementation of the above equidistance principle cannot be applied. In these cases, the user will find the capabilities provided by the Utility Programs and Offshore Limit modules most useful.

Weighted boundaries

The Weighted Boundary module is formed by two programs: *Two Step* and *Equiratio*. Two Step follows the traditional approach of determining a boundary in which one or several points have been assigned 'partial effects'. The presence or absence of a geographic feature are assumed consecutively, and the equidistant boundary is determined in both cases. The resulting boundaries are then used to determine a 'weighted boundary'. *Equiratio* is an implementation on the surface of a geodetic reference ellipsoid of the method developed by Langeraar (1985, 1986) for island states. It determines a boundary as a conic section over a user-prescribed region from two points belonging to two opposite coastal states. The type of conic section is controlled by the selected ratio of the distances from the two points to the boundary.

Utility programs

The Utility Programs module is formed by seven programs: *Direct, Inverse, Fill, Sweep, Grid, LatLon* and *Trans*.

Direct and *Inverse* solve the direct and inverse positioning problems on the surface of a geodetic reference ellipsoid. These problems are defined as follows:

(a) given the latitude and longitude of one point and a distance along a geodesic azimuth to a second point, find the latitude and longitude of the second point; and

(b) given the latitude and longitude of two points, find the distance along the geodesic line joining them and the azimuth of the geodesic at both points.

Geodesics over the ellipsoid are selected because by definition they avoid any geometric ambiguity. A geodesic is the curve C over which the distance S between two points is a minimum:

$$S = \min \int_c dS$$

Fill allows the user to determine a user-prescribed number of equally spaced points along a segment of a geodesic defined by its two end points. This application is of interest in at least two instances during the use of DELMAR. First, it describes straight or archipelagic baselines (Prescott 1987) along the coast by segments of geodesics, and second, it determines the coordinates of any number of intermediate points between boundary turning points found formerly by other programs, such as Equid-23.

Sweep uses the output information of any options on the Offshore Limit module to determine a user-prescribed number of points in a counter-clockwise direction along an arc limited by two offshore points. These offshore points can be located at 12,200 or a user-prescribed number of nautical miles from a known point.

Grid and *LatLon* may be needed if, instead of performing a geodetic survey with the purpose of determining the coordinates of coastal points, positions are obtained through the use of a nautical chart and a digitiser. *Grid* makes it possible to transform values generated in centimetres or inches by a digitiser into northings and eastings on a grid of a Mercator mapping plane. *LatLon* performs a similar task but transforms digitiser values into geodetic latitude and longitude.

Trans is used to transform geodetic coordinates from one reference ellipsoid to another, provided that their transformation parameters – three rotations, three translations, and their semi-major and semi-minor axes – are available.

Text editor

The *Text Editor* program provides a full-screen text editor with which the user can create, view, or modify input and output data files. It generates ASCII files that can be used by DELMAR and can also be sent to a printer if hard copies are desired.

Graphics screens

The Screen Graphics module is formed by four programs: *CGA*, *HERCULES*, *EGA*, and *VGA*. These programs support the four most popular graphics adaptors installed in personal computer systems around the world.

Support programs

A computer program that is invisible to the user but nevertheless performs an important task also runs as a part of DELMAR. This program is *RDEM02.EXE* by Bricklin (1987). It uses TUTOR.DBD to run the tutorial. RDEM02.EXE is a copyrighted program, but the terms of its license agreement allow it to appear in DELMAR.

LANGUAGE AVAILABILITY

A special effort has been made to develop several international versions of DELMAR. Equally functional versions of the software and printed manuals are available in English, French and Spanish.

CONCLUSIONS

DELMAR is a computer library that has filled a very important gap in the technical literature of Law of the Sea. Version 1.0 of this computer library was designed to carry out several of the most frequently demanded technical tasks in the delimitation of maritime boundaries and offshore limits. DELMAR has found wide acceptance worldwide as documented by its present use in over forty different countries. Feedback from all users will prove extremely valuable for the development of a greatly enhanced version 2.0 scheduled for production in the last quarter of 1992.

REFERENCES

Beazley, P.B. (1978) 'Maritime limits and baselines', *The Hydrographic Society Special Publication No. 2*, Second edition.
Blake, G. (ed.) (1987) *Maritime Boundaries and Ocean Resources*, London: Croom Helm.
Bricklin, D. (1987) *Dan Bricklin's Demo II Program User Manual*, Boston: Software Garden.
Carrera, G. (1987) 'A method for the delimitation of an equidistant boundary between coastal states on the surface of a geodetic ellipsoid', *International Hydrographic Review* LXIV, 1, 147–59.
Duncan, R. (1986) *Advanced MS-DOS*, Washington: Microsoft Press.
Jagota, S.P. (1985) *Maritime Boundary*, Dordrecht: Martinus Nijhoff Publishers.
Johnston, D.M. (1987) *The Theory and History of Ocean Boundary Making*, Kingston: McGill–Queens University Press.
Langeraar, W. (1985) 'Equitable apportionment of maritime areas through the equiratio method', *Hydrographic Journal* 36, 19–28.

Langeraar, W. (1986) 'Maritime delimitation. The equiratio method – a new approach', *Marine Policy* 10, 1, 3–18.

Norton, P. (1985) *PC-DOS Introduction to High Performance Computing*, New Jersey: Prentice Hall.

Prescott, J.R.V. (1985) *The Maritime Political Boundaries of the World*, London: Methuen.

Prescott, J.R.V. (1987) 'Straight and archipelagic baselines' in G. Blake (ed.), *Maritime Boundaries and Ocean Resources*, London: Croom Helm.

Shalowitz, A.L. (1962) *Shore and Sea Boundaries*, Washington DC: US Department of Commerce, 2 vols.

United Nations (1953) Document A/CN.4/61/Add. 1.

United Nations (1982) *Convention of the Law of the Sea*, Conference Document A/Conf.62/122.

United States Department of State (1970–1987) *Limits in the Seas*, Washington DC: Office of the Geographer.

Vanicek, P. and E. Krakiwsky (1986) *Geodesy: The Concepts*, Amsterdam: North Holland.

5

CORAL REEFS AND THE 1982 CONVENTION ON THE LAW OF THE SEA[1]

Peter B. Beazley

Before 1973 no special consideration was given to the status of drying coral reefs as baselines for the territorial sea, but even before the Third United Nations Conference on the Law of the Sea atolls were perceived as being features the importance of which was inadequately recognised under the current law. To remedy this, Article 6 of the 1982 Convention was drafted. References to reefs also appeared in Article 47 dealing with archipelagos.

Art. 6 Reefs
In the case of islands situated on atolls or of islands having fringing reefs, the baseline for measuring the breadth of the territorial sea is the seaward low-water line of the reef, as shown by the appropriate symbol on charts officially recognized by the coastal State.

Art. 47 Archipelagic baselines
1. An archipelagic State may draw straight archipelagic baselines joining the outermost points of the outermost islands and drying reefs of the archipelago ...
7. For the purpose of computing the ratio of water to land ... land areas may include waters lying within the fringing reefs of islands and atolls, including that part of a steep-sided oceanic plateau which is enclosed or nearly enclosed by a chain of limestone islands and drying reefs lying on the perimeter of the plateau.

These provisions are distinct from Article 13 dealing with low-tide elevations, and whereas application of that Article is optional, Article 6 applies *ipso jure*.

Article 6 originates from a paragraph in an article on islands submitted to UNCLOS III in 1974 by Fiji, New Zealand, Tonga and

Western Samoa. It was introduced by the New Zealand representative who stated that the draft was 'designed to fill a gap in the existing law concerning baselines for the territorial sea as that law applies to atolls and other island systems with the same features as atolls'. He also pointed out that the lagoon waters constituted the principal source of food for the inhabitants of an atoll.

THE PHYSICAL FEATURES OF REEFS

The term 'reef' refers to a mass of rock or coral which either reaches close to the sea surface or is exposed at low tide. That part of a reef which is above water at low tide but submerged at high tide is a 'drying reef'.[2] Coral reefs are tropical shallow water ecosystems, and may be found in most areas of the tropics, but they are most prevalent in the central and western Pacific Ocean. They take four main forms: atolls, fringing reefs, barrier reefs, and patch reefs. Patch reefs form on irregularities on shallow parts of the seabed, and may occur inside atoll lagoons. An atoll is a ring-shaped, but sometimes elliptical or horseshoe-shaped, coral reef which has islands or islets on it, the shallow rim enclosing a deeper central area or lagoon[3] (Figure 5.1). Not all authorities insist on there being islands on an atoll.[4]

Maloelap is a typical atoll. As with all coral reefs the main growth is on the windward side, where the reef tends to be continuous. It has numerous small islands on it. As the reef grows it leaves dead coral behind it to produce a large flat surface which is quite unlike a wave-cut rock reef. Usually there are passages through the reefs on the leeward side of the atoll. The perimeter cover is generally greater than 75 per cent, and usually averages 90 per cent.

To geomorphologists a fringing reef is a coral reef that is directly attached to or borders the shore of an island or continent, having a rough table-like surface that is exposed at low tide. There may be a shallow channel or lagoon between the reef and the land upon which it is attached.[5] A fringing reef in fact would be the normal low water line as shown on a nautical chart. Generally the boat channel would not show on the scale of the chart. Neither would an island having only a fringing reef be in any way atoll-like.

A barrier reef is a long and narrow coral reef lying roughly parallel to the shore but separated from it by a wide and deep lagoon. Islands often form on barrier reefs just as they do on atoll reefs. Uvea is a very good example of an atoll-like feature, with both barrier and fringing reefs (Figure 5.2). The chart shows that there is an area of fringing reef on the

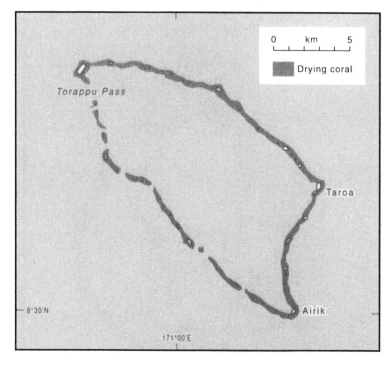

Figure 5.1 Maloelap in the Marshall Islands: a good example of an atoll

west coast which projects to merge with the barrier reef. This feature of fringe merging into barrier is quite common, making it difficult to distinguish the two.[6] Writers are generally agreed that the reference in the Convention to 'fringing reef' must include barrier reefs that form a fringe off an island, so that in effect the term 'fringing' is used somewhat as in Article 7 to refer to islands forming a fringe along a coast.[7]

There are two examples of terms which have a juridical meaning narrower or wider than ordinary meaning: Article 10 dealing with bays, and Article 76 dealing with the continental shelf. The first limits the type of feature that may be considered as a juridical bay, and the second vastly extends the area of seabed that may be considered as juridical continental shelf. One should not be surprised, therefore, if the term 'fringing' is also used in a special juridical manner.

Some authorities say the term 'fringing' can also apply to ordinary rock reefs.[8] This is not, however, significant in the light of the intent of Article 6. A true fringe is irrelevant since it is the low water line of the

61

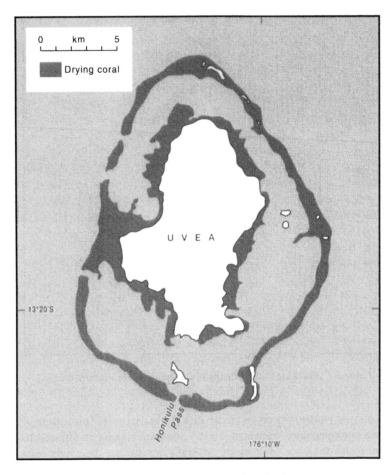

Figure 5.2 Uvea in the Wallis Islands

island anyway. I know of no case where a barrier reef or ordinary rock reef exists to product an atoll-like situation. From the authors of the Article, and its intent, there can be no doubt that the term reef refers specifically to a coral reef.

CHARTING PRACTICE

A characteristic of coral reefs is the very large area of flat reef at about the level of low water. Corals cannot live or grow out of water, so quite large areas of the surface of a 'drying' coral reef may lie at or just below

low water level, but with pinnacles of coral or extensive patches of debris and sand which rise above it.

When a hydrographic survey is made of a coral reef preparatory to making a nautical chart, it is generally too dangerous to approach close to the reef from seaward, and the reef is often too shallow to navigate safely even in a small sounding boat. No attempt may be made to determine exactly what dries and what doesn't. The whole area of the flat is designated 'drying reef' and outlined by the appropriate symbol for a coral reef which dries at low water. The extent to which individual reefs or parts of reefs are shown as distinct features depends, as is the case with other charted features, on the scale of the chart. The larger the scale of the chart the greater the discrimination.

From about the middle of the nineteenth century all states which published nautical charts have used a distinctive symbol to indicate areas of rock or reef which uncover at low tide. Since the 1930s several countries have adopted a special symbol for drying coral reefs as distinct from rock reefs, whilst others have been content to mark the distinction by an abbreviation such as 'Co'. With the introduction of the *International Chart* series a symbol has been agreed which is being adopted by many countries. But no matter when a chart was first published, if it depicts drying coral reefs it will do so by the use of an 'appropriate symbol' from which, with or without the aid of an abbreviation, it is possible to determine that the area is coral.

Some countries used to show the drying symbol out to the point where the reef plunged to deep water. However, it is the seaward edge of the reef that matters for baseline purposes. On that edge, which is very steep, the horizontal difference between the low water line and the plunge line would be undiscernible.

LEGISLATIVE HISTORY

During the nineteenth century the 'portico doctrine', by which small offshore features were accepted as territorial sea, eventually led in some cases to the inclusion as territorial sea baselines of features such as shoals and reefs.[9] The term reef without qualification, however, does not necessarily refer to a coral reef, neither is it often distinguished in treatment from rocks, banks or shoals. Thus the 1893 Russian claim to measure from the farthest islands, rocks, banks of stone or reefs showing above the sea[10] was certainly not referring to coral reefs.

It was not until 1953 that any special consideration was given to coral features. The International Law Commission's committee of experts on technical questions concerning the territorial sea was asked: if in principle the territorial sea was to be measured from the low water line, what line should be adopted as such? They answered that in the case of coral reefs the edge of the reef as shown on charts should be considered as the low water line for drawing the limit of the territorial sea.[11] However, no mention of coral appears in the 1958 Convention.

In 1973 Malta submitted to the UN Seabed Committee draft articles one of which made special provision for atolls.[12] It was largely ignored and the concepts outlined in it were not repeated in the 1974 proposals.

STATE PRACTICE

In 1862 – during the American Civil War – the Law Officers were asked a question about the territorial extent of Bermuda. They said that North Rock permitted of a fort being constructed on it and it was unthinkable that it should not be within Bermudan control. The reefs generally, though covered at high water, are a natural ledge or girdle of defence. Territorial waters should be measured from the outer edge of the reefs, which the Law Officers thought uncovered at low water[13] (Figure 5.3). This was perpetuated by the Interpretation Act of 1951 and, more recently, in 1988, by an Order in Council.

Despite the example of the Law Officers' Opinion concerning Bermuda, the UK from about 1880 resolutely refused, in any debate about territorial sea measurement, to accept that territorial claims could arise from any offshore feature other than habitable islands. In its pleadings in the Anglo-Norwegian Fisheries Case the UK claimed that the Law Officers could not have consulted the chart.

A 1911 French decree stated that the territorial sea of New Caledonia was to be measured from 'the great outer reefs, and where there are no reefs, . . . from the low-water line'.[14] The UK's actions show that, whatever her formal views, she was prepared to be pragmatic about territory in remote areas with widely scattered small insular features far from international shipping routes. The 1942 Fiji Fisheries Ordinance enacted that:

'territorial waters' means that part of the sea adjacent to the coast of any island in the Colony which is within three geographical miles measured from low-water mark of the seaward side of the reef fronting such coast, or when a reef is not present, from the low-water mark of the coast itself.

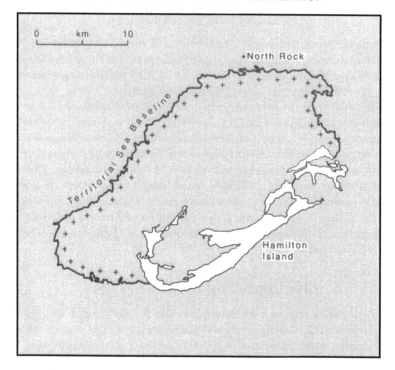

Figure 5.3 Bermuda Islands: the chart shows no area of drying reef

In 1946 a somewhat similar Ordinance was enacted for the Gilbert and Ellice Islands (now the independent states of Kiribati and Tuvalu):

> 'territorial waters' means that part of the sea adjacent to the coast of any island in the Colony which is within three geographical miles measured from low-water mark of the seaward side of the reef fronting such coast or bounding any lagoon waters adjacent to such coast, or, when a reef is not present, from the low-water mark of the coast itself.

Although the latter is slightly more explicit in referring to reefs bounding a lagoon, neither ordinance places any restriction on the distance of the reefs from the islands.

By 1972 independence, and the increasing interest in tuna fishing and seabed nodules, brought a demand from the now numerous insular states for something that was accepted by international treaty, and could allow sovereign control over the fragile ecology of reefs and the lagoons which depended on them.

APPLICATION OF ARTICLE 6

Article 6 applies *ipso jure* – it refers only to drying reefs – there is no limit on distance – and, strangely, there is no provision for closing lines across openings in the reefs. Writers and such state practice as is known are generally agreed that closing lines may be used.

It could hardly be otherwise. The baseline is to be the *seaward low-water line* of the reef. Thus at a reef opening the baseline apparently ceases; it cannot follow the reef around into the lagoon, because it would cease to be the seaward edge. Reefs are the baseline where they front islands; the islands themselves do not form the baseline. Thus unless the openings are artificially closed there is no complete closure and no protection for the lagoon waters; that would defeat the purpose of the Article. A further point is that the import of Article 8 is that the waters behind the reefs should be internal waters, and they could not be so without closing lines.

DISCUSSION OF ALTERNATIVES

The pragmatic approach by those states which have introduced legislation specifically dealing with atolls and fringing reefs has ensured that deficiencies in that Article, which defines neither atoll nor fringing reef, and fails to provide for the complete enclosure of lagoon waters, have not yet resulted in any obvious breach of its spirit.

The Convention is not yet in force, and it is interesting to consider whether the intent of Article 6 could not have been accomplished just as well through the provisions of Article 11 of the 1958 Convention, which is the same as article 13 of the 1982 Convention.

Article 13: low-tide elevations

1. A low-tide elevation is a naturally formed area of land which is surrounded by and above water at low tide but submerged at high tide. Where a low-tide elevation is situated wholly or partly at a distance not exceeding the breadth of the territorial sea from the mainland or an island, the low-water line on that elevation may be used as the baseline for measuring the breadth of the territorial sea.

So long as a number of major maritime nations refused to accept the legitimacy of claims to a 12-mile territorial sea, this article could not with certainty embrace within the baselines all the reefs comprising the limits of the lagoon waters of atolls or similar features. A 12-mile limit, however, does encompass such reefs in all but exceptional cases. If all the drying reefs on the perimeter of the atoll lie wholly or partly within 12 miles of the low water line of the islands, then they will all be base-lines, and to that extent Article 6 in its present form is unnecessary.

It is a characteristic of Article 13, as shown by the *travaux préparatoires* to the 1958 Convention, that the low-tide elevations concerned are in effect to be treated as if they were islands. Thus the reefs would be assimilated to islands, and the waters between the reefs and the actual islands would not be internal unless they could be claimed as such on historic grounds – a not improbable claim. On the other hand, the lagoon waters of such features are often difficult of access, and the usable passages tend to be situated on one side only of the lagoon, which itself will tend to be reef-encumbered and in-adequately surveyed for navigational purposes. In view of that and the resulting dangers to the coastal state from pollution arising from grounding, it would be difficult to claim that there must be a right of innocent passage through the lagoon from one part of the high seas to another when an unencumbered passage is available outside the reefs. No prudent mariner would consider making such a claim. None of the atolls or islands suitable to be dealt with under Article 13 are so large or so proximate to one another as to make passage through the deep water outside the reefs a diversion of a vessel's route.

Article 47(1)

The term 'reef' is also used in Paragraph 1 of Article 47. The reference to drying reefs was in the first draft of archipelagos submitted to the Seabed Committee by Fiji, Indonesia, Mauritius, and Philippines in 1972.[15] This draft preceded the prototype draft for Article 6 but, like that Article, was submitted by states all of which are in localities where 'reef' is synonymous with 'coral reef'.

Article 6 which is entitled 'reefs' has been shown to be concerned solely with coral reefs. These are unique in that drying coral reefs are often of considerable extent and possess a large flat surface, unlike a typical detached wave-cut reef. The large emergent surface of drying coral reefs, and the materials available from the sheltered lagoon waters that may be present, has made them features which, with modern means of access and modern technology, may be capable of development into artificial islands.

There are non-coral reefs on which lighthouses have been built but this writer can think of no rock reef, located outside what would otherwise be territorial sea and with no part of it permanently above water, of sufficient extent to be suitable for development as an artificial island in the manner considered in the preceding paragraph. The provenance of the reference to drying reefs in Article 47(1), the fact that Article 6 refers only to coral reefs, and the greater security threat posed by it, support the view that 'reefs' here also refers to coral reefs. Paragraph 4 permits the use of low-tide elevations subject to certain conditions. If drying reefs are to be considered as no more than low-tide elevations subject to the conditions imposed by Paragraph 4, it would be expected that the term 'low-tide elevations' would have been used instead of 'drying reefs' in Paragraph 1. Thus Paragraph 1 would permit the use of low-tide elevations and Paragraph 4 would be more specific as to the conditions in which they may be used. In fact, though, just as in the case of Article 5, reefs are to be viewed as being more than just low-tide elevations and are to be given special treatment accordingly, which would be entirely consistent with the special treatment of certain types of coral reef in the earlier article.

At the second session of the Conference at Caracas in 1974 a working paper submitted by Canada, Chile, Iceland, India, Indonesia, Mauritius, Mexico, New Zealand, and Norway included drying reefs without qualification.[16] At the same session the Bahamas submitted draft articles which included a paragraph:

> 1. In drawing the baselines ... an archipelagic State may employ the method of straight baselines joining the outermost points of the outermost islands and drying reefs *or low-tide elevations of the archipelago* or may employ any non-navigable continuous reefs or shoals lying between such points (emphasis added).[17]

This draft clearly distinguishes between drying reefs and low-tide elevations. Given the security implications that arise from the existence of an emergent coral reef within the geographical unity of an archipelago, there is certainly need to include it within the archipelagic waters. Another practical consideration is that, if the drying reefs of an atoll, or of an island with fringing reefs, forming part of the archipelago, lie more than 12 miles from the low water line of the islands or island, they would be baselines under Article 6. It would therefore be illogical for them to be excluded as archipelagic basepoints. Similarly it could not be the intention that in applying Article 47(7) to determine the water to land ratio, some of the fringing reefs of islands and atolls might lie outside the archipelagic baselines.

The conclusion is inescapable, that the inclusion of drying (coral) reefs as basepoints is not to be limited by the provisions of Paragraph 4, but only by Articles 46(b) and by Paragraphs 1, 2, 3 and 5 of Article 47.

Article 47(7)

Paragraph 7 does not apply *ipso jure*. Prima facie it is intended to apply Article 6 in such a way as to allow the internal waters of Article 6 reefs to be included as land for the purpose of determining the water to land ratio. This would be entirely consistent with the approach to such internal waters as lakes and rivers. Indeed, at first sight it appears to repeat much of the wording of Article 6, but there are three important differences. First, there is no mention of 'drying' reefs; second, there is no reference to islands on atolls; third, there is an additional feature which is not mentioned in Article 6 at all. This is the reference to a steep-sided oceanic plateau etc., added in order to accommodate the concerns of such countries as the Bahamas.[18] These differences destroy the superficial mirroring of Article 6.

As a result the paragraph is open to the interpretation that an archipelagic state may use submerged reefs as a boundary to an area that it wishes to include as land for the purposes of determining the water to land ratio. This interpretation appears to have been adopted by Fiji in practice. On Fiji Marine Spaces Chart 81/2, prepared to illustrate the application of the Marine Spaces Acts of 1977 and 1978, the limit of internal waters off the western end of Vanua Levu are shown as limited by a line following the outer edge of a shallow bank extending about 45 miles from the island. This bank is edged by drying reefs on the north side, but to the west and south the reefs are submerged (Figure 5.4).

There are definitions of an atoll that omit reference to there necessarily being islands associated with the reefs. The omission of a reference to islands in Paragraph 7 may, therefore, be significant. The Maldives consist of numerous atolls, some of which are compound atolls, on which there is no surrounding drying reef forming a barrier to enclose a lagoon. Instead the rims of the atolls have numerous faros, or small atolls, strung along them. From the charts it is impossible to see whether they all have small islands upon them, but it appears that many have not.

Compound atolls of this sort do not easily fall to be treated under Article 6 since there is no clear encircling reef, although it is possible to draw a series of straight baselines linking the faros and individual islands around the perimeter. But the 'lagoon' waters of these atolls,

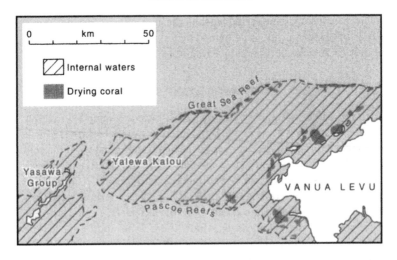

Figure 5.4 Part of Fiji Islands

some of which stretch over considerable distances, are not in any way comparable to those of smaller and more enclosed atolls. They are not difficult of entry, and there are large areas where it would appear to be possible to navigate a vessel, albeit with caution in the absence of a modern chart.

An alternative to considering these features under Article 6 is to consider them as archipelagos. If the faros are considered as land, even if they have no islands upon them, the water to land ratio could probably be satisfied in most cases.

The third element of this paragraph, that which refers to a steep-sided oceanic plateau, does in fact describe an atoll, and unlike the other elements of the paragraph it requires that the feature should be enclosed or nearly enclosed by a chain of limestone islands and *drying* reefs. It can hardly be held to mean that the term 'atoll' in the same paragraph does not mean atoll at all. It is necessary, therefore, to consider the feature or features to which it might have been intended to refer. Looking specifically at the Bahamas, the most likely – possibly the only – area which would match the description is Grand Bahama and Great Abaco Islands with their associated islets and reefs lying on Little Bahama Bank. The whole bank covers an area about 140 nm long by about 50 across at its widest point. Although at one time thought to be an atoll formation, that interpretation is no longer tenable.

Although Article 6 is intended to include 'other islands with the same

70

features as atolls', it is doubtful whether, in formulating that Article, thought was given to a formation of the size of the Little Bahama Bank. It therefore seems that it is to take account of a very large feature of that sort that the definition was included in Article 47.

CONCLUSIONS

Article 6 refers to the drying reefs of atolls and of other atoll-like islands, and is confined to coral reefs. The term 'fringing reef' is ambiguous, and often there is no clear distinction between fringing and barrier reef. The baseline of an island with a true fringing reef is the low-water line of the reef under the provisions of Article 5. Fringing reef in Article 6 refers rather to reefs forming a fringe around an island.

The article lacks clarity, however, and makes no provision for gaps in the reefs, although this has been treated pragmatically in practice. Nevertheless, it probably achieves nothing that could not be effected under the provisions of Article 13, which, as Article 11 of the 1958 Convention, may have greater legal force. The pragmatic way in which real geographical conditions have been treated suggests that in most cases the question is academic, and unlikely to give rise to dispute. One cannot be so sanguine about the interpretation of the relevant paragraphs of Article 47.

Paragraph 1 of Article 47 also refers to coral reefs, although it is not limited to fringing/barrier reefs. Neither is it governed by Paragraph 4, but only by the other provisions which refer to the islands of the archipelago and their enclosure by archipelagic baselines. The first part of Paragraph 7 Article 47 cannot be read as referring only to geographical situations covered by Article 6. There are significant differences both as regards atolls and as regards the question whether the reefs are drying reefs. The reference to a steep-sided oceanic plateau, being intended simply as a definition, appears in its present context to be rather enigmatic. The reason for its inclusion, however, provides a strong pointer to its intended interpretation.

NOTES

1 This paper is based on the results of a study undertaken at Hull University. It is an abridged version of a paper 'Reefs and the 1982 Convention on the Law of the Sea' published in the *International Journal of Estuarine and Coastal Law*, vol. 6(4), 1991, pp. 281–312.

2 *A Manual on Technical Aspects of the United Nations Convention on the Law of the Sea – 1982*, International Hydrographic Organization Special

Publication No. 51 (IHO SP51), Monaco 1990, p. 24.

3 See e.g. Susan M. Wells (ed.), *Coral reefs of the World*, vol. 1, Cambridge, 1988, p. xvi and Gary, McAfee and Wolf, *Glossary of Geology*, Washington DC, 1972.

4 For example 'A ring-shaped reef with or without an island situated on it surrounded by the open sea, that encloses or nearly encloses an island', IHO SP51, op. cit., p. 8.

5 See *Glossary of Geology*.

6 See S. Gardiner, *Coral Reefs and Atolls*, London, 1931, p. 13.

7 UN *Baselines*, UN Office for Ocean Affairs and the Law of the Sea, New York: UN, 1989, No. E88V5*, p. 10; P.B. Beazley, *Maritime Limits and Baselines*, 3rd edn, Hydrographic Society SP3, Dagenham 1987, p. 12; H.W. Jaywardene, *The Regime of Islands in International Law*, Dordrecht, 1990, p. 99.

8 See IHO SP51, op. cit., p. 8 or *Mariners Handbook* (6th edn.), HMSO, 1989.

9 See D.P. O'Connell, *The International Law of the Sea*, vol. 1, Oxford, 1982, p. 185.

10 See P. Fauchille (1925), cited by J.R.V. Prescott, *Australia's Maritime Boundaries*, Canberra, 1985, p. 9.

11 UN Doc. A/CN.461/ADD, 1 of 1953; *ICL Yearbook 1953*, vol. II, pp. 75–9.

12 UN Doc. A/AC.138/SC.II/L.28. S. Oda, *The International Law of Ocean Development*, vol. II, Leiden, 1975.

13 H.A. Smith, *Great Britain and the Law of Nations*, vol. II, London, 1935, pp. 233, 234.

14 Supplied by courtesy of the French Service Hydrographique et Océanographique de la Marine, supra n. 18.

15 Originally issued as document A/AC.138/SC.II/L/15. See UN 'Law of the Sea' publication (Archipelagic States), UN Sales No. E.90.V.2, n. 21.16.

16 Document A/CONF.62/C.2/L.4.

17 Document A/CONF.62/C.2/L.70.

18 'The second half of the paragraph, starting with "including", was added in order to accommodate the concern of such countries as the Bahamas.' See UN *Baselines*, p. 78.

6

SEA LEVEL RISE AND MARITIME BOUNDARIES
International implications of impacts and responses

David Freestone and John Pethick

Sea level rise is one of the more widely publicised possible consequences of global warming: for coastal areas it will certainly be one of its most important impacts. Working Group 1 of the Intergovernmental Panel on Climate Change (IPCC, WG1 1990) now suggests that the most reliable models project a 'best estimate' of an 18-cm rise caused principally by the melting of small glaciers and the general increase in the mass of the sea induced by warming. Thereafter, if there are no limitations on the melting of small glaciers and the general increase in the mass of the sea induced by warming. Thereafter, if there are no limitations on the emission of greenhouse gases (the 'business as usual' scenario) the models suggest that the planet will be committed to a steep rise in temperature and continuing and increasing rises in sea level.[1]

Nearly 65 per cent of the world's population lives within a few kilometres of the coast. Although present standards of contour mapping (which omit, for example, both 1 m and 50 cm lines) do not allow accurate assessment of the extent of coastal areas which would be affected by such a rise, the IPCC Coastal Zone Management Subgroup (1990) reports that some 15–17% of Bangladesh, 17% of the arable land of Egypt, the Mekong Delta, and many of the coral atolls of the Maldives would become uninhabitable after a rise of one metre. So too would Cocos Island, and the atolls of Tokelau, Tuvalu, Kiribati, the Marshall Islands and the Torres Strait Islands in the Pacific.

Higher sea levels would also have an effect on the dynamics of the ocean system. There would be changes in currents and thus the patterns of upwellings and fish abundance. Some argue that warming itself will also affect the incidence and severity of storms and hurricanes and the higher ocean levels might therefore bring disproportionate increases

73

in the severity of storm-related flooding. It should be clear, therefore, that if these current predictions are correct, sea level rise threatens to cause major impacts on the shorelines. This will have considerable significance for boundaries, in that such changes will affect the physical configuration of the coast, both in terms of the geographical features from which maritime zones are measured and the actual physical boundary between the land and the sea. This paper seeks to explore some of the implications that such a projected rise in sea level would have on international maritime boundaries and the implications of possible policy responses which may be adopted by coastal states.

The first and most obvious effect which rises in sea level will have is on high and low water lines. Changes in the coastal baseline, from which national maritime zones are measured, will thus have an effect on the maritime areas which a state may legitimately claim and on its jurisdiction over the exploration and exploitation of resources, including fisheries and minerals, within those zones.

COASTAL BASELINES

Article 5 of the 1982 Law of the Sea Convention (LOSC), which is taken to represent customary international law, provides that 'the normal base line for measuring the breadth of the territorial sea is the low water line along the coast as marked on large scale charts officially recognised by the coastal state'.

STRAIGHT BASELINES

In circumstances specified by Article 7 (LOSC), coastal states are allowed to use straight baselines joining headlands, islands, rocks, and other features. In crude terms, rises, or indeed any changes, in sea level which affect high and low water lines will obviously have a 'knock-on' effect on the measurement of maritime jurisdictional zones. Where a broad coastline regresses rapidly (as in the case of areas of Bangladesh which are being eroded at up to 140 metres a year (Stoddart and Pethick 1984)) the cumulative effect can be quite substantial.

In fact Article 7(2) (LOSC) does contain a provision – derived from the proposal initiated by Bangladesh which was concerned by the particular problems of constant erosion and deposition at the mouth of the Bramaputra – permitting in restricted circumstances a straight baseline to be maintained notwithstanding the movement of the actual coast:

74

Where because of the presence of a delta and other natural conditions the coastline is highly unstable, the appropriate points (i.e. for straight baselines) may be selected along the furthest seaward extent of the low-water line and, notwithstanding the subsequent regression [*sic*] of the low-water line, the straight baseline shall remain effective until changed by the coastal state in accordance with this Convention.

Although, as Prescott and Bird point out, this was drafted for specific circumstances, there is a risk that in the context of sea level rise this limited exception will be utilised more widely than is legitimate under the LOSC regime (Prescott and Bird 1990).

LOSS OF BASEPOINTS

Article 7 (LOSC) permits a coastal state to use straight baselines 'in localities where the coastline is deeply indented and cut into, or if there is a fringe of islands along the coast in the immediate vicinity'. However, straight baselines may only be drawn from 'appropriate points', namely islands and other features which are above water at high tide; low-tide elevations may only be used as basepoints if 'lighthouses or similar installations which are permanently above sea level have been built on them'.

If a low water line may be maintained by extensive, and of course expensive, artificial means, then surely similar artificial means may also be used legitimately to maintain straight baseline points by the artificial maintenance of islets, rocks and other similar features, to prevent them becoming low-tide elevations. Thus, new artificial installations may be built on such new low-tide elevations to bring them within the term of Article 7(4). Indeed, as Prescott and Brid (1990) point out, coastal states are in any event likely to use the further proviso of Article 7 that the drawing of baselines to and from low-tide elevations is permissible if it 'has received general international recognition' to argue that features used on charts and subsequently reduced or submerged by sea level rise have already received such recognition.

ISLANDS

A similar problem is posed by islands themselves. Under Article 121(1) an 'island', i.e. 'a naturally formed area of land surrounded by water', is entitled to claim the full suite of maritime zones. This is however subject to the important qualification in Article 121(3) that 'Rocks which cannot sustain human habitation or economic life of their own shall have no exclusive economic zone or continental shelf.'

Although considerable controversy surrounds this latter provision (Kwiatkowska and Soons 1990), there is at least prima-facie evidence that it has been recognised as customary international law (Freestone 1990). Whatever the merits of that debate, sea level rise might well cause low-lying islands to disappear entirely, or to lose so much of their area or their fresh water 'lens' as to become uninhabitable and thus be transformed into 'rocks' with the consequent loss of entitlement to maritime zones previously established.

Professor Soons, in his pioneering paper (Soons 1990), has explored a number of options open to a coastal state facing such a situation. The weight of doctrinal opinion suggests that the artificial enhancement of an existing 'rock' so as to make it habitable or capable of sustaining an economic life of its own does not transform it into an island, but the artificial maintenance of an island to prevent it becoming a 'rock' for the purposes of Article 121(3) might, Soons thinks, be permissible. Certainly, the downgrading of a feature from an 'islet' under the terms of Article 121(1) to a 'rock' under Article 121(3) would pose considerable difficulties in determining the status of the maritime areas which it had previously generated and which may have been recognised by other states as being under the sovereignty or jurisdiction of the state to which it belongs.

Some states may seek to utilise the doctrine of historic waters but in view of some of the dangers discussed below, there is considerable merit in Soons' proposal (Soons 1990) for the development of a new concept of customary law that:

> Coastal states are entitled, in the case of landward shifting of the baseline as a result of sea level rise, to maintain the outer limits of the territorial sea and of the EEZ where they were located at a certain moment in accordance with the general rules in force at that time.

Indeed such a rule would be useful to include within a Sea Level Rise Protocol to the Climate Convention – as proposed by the IPCC Coastal Zone Management Subgroup (IPCC/CZM, 1990).

CONTINENTAL SHELF

Baseline movement will not normally affect the limit of continental shelf claims which may extend to the edge of a continental margin. It could however have important economic effects on equidistance lines which may be significant in areas where boundaries have still to be settled, albeit by reaching an 'equitable solution', in that hydrocarbon resources may move over a median line. The inundation or reduction of small islets or features which are currently used or are in dispute as basepoints for maritime zones will also have effects on maritime boundary negotiations. Although the coral around coral-based islands might be able to keep pace with a regular increase in sea level, it is more difficult to predict whether the island could renourish itself from that reef within a predictable time scale so that it might disappear entirely (perhaps to reappear in the next century). Soons takes the view that the artificial conservation of such features also 'should be permissible' (Soons 1990).

Because of the complexity of the system established by Article 76 of the 1982 LOS Convention for the calculation of the outer limits of the continental shelves of coastal states, the Convention envisages a Commission on the Outer Limits of the Continental Shelf which would assess the conformity with the requirements of the Convention of continental shelf claims submitted to it. Outer shelf limits established by the coastal state on the basis of the recommendations of this Commission 'shall be final and binding'. Although the primary objective of this procedure was to provide a clear boundary between coastal state authority and that of the Sea Bed Authority, a possibly unlooked-for result of this 'permanent fixing' may be, as Soons points out, that:

> Once the outer limit of the Continental Shelf has been established at a distance of 200 nautical miles (or even more) from an island, which island subsequently disappears entirely, the coastal state would maintain sovereign rights over a seabed area (which may be of considerable extent) while the object which generated the sovereign rights no longer exists.
>
> (Soons 1990)

Strange though this result may be, it may provide an unexpected impetus to states both to ratify the LOS Convention and to submit their claims to the Commission for assessment.

EFFECT ON MARITIME BOUNDARIES

Maritime boundary agreements, once made, belong to that class of treaty the validity of which is not affected by 'subsequent fundamental change of circumstances' under the Vienna Convention on the Law of

Treaties 1969 (Article 62(2)(a)). However, some maritime boundary treaties do utilise moving concepts such as the *thalweg* (in rivers), the line of greatest depth or strongest current or the centre of the main navigable channel, in maritime law the centre of the principal navigable channel (O'Connell 1984) which may be considerably affected.

Rises of over a metre would have major impacts on coastal currents. In areas yet to be determined, changes in baselines could have a significant effect on the negotiating position of the parties – which are still influenced by equidistance lines – even though enjoined by the LOSC (Article 83) simply to reach an 'equitable solution'. However it seems clear that in the majority of areas, high and low water lines will not necessarily move inland at the same pace. Depending on the geomorphology of the coast, the tidal range might decrease or even increase. In the latter case of course, regression of the low water line would have the opposite effect to that described above.

POLICY OPTIONS

The delineation of baselines is primarily a matter for the coastal state (see Article 5 (LOSC) above). Prescott and Bird have suggested that this might present two policy options: the active option – continuous updating of charts; and the passive option – leaving the charts alone (Prescott and Bird 1990). Apart from the expense of the active option, which is significant particularly for small developing countries (who might be relatively the most affected), there is a problem that such action might be seen as the unilateral abrogation of existing maritime areas and hence politically undesirable. The passive option however is dangerous. International law simply requires that the low water line be marked on 'large scale charts officially recognised by the coastal state' (Article 5 (LOSC)). There is no requirement that these be specifically produced for baseline delineation and indeed they are not. The charts used are designed primarily for navigation. Hence, charts left unchanged for political reasons at a time when important coastal geomorphological changes are underway could be extremely hazardous. Or we could see the widespread evolution of baseline maps (similar to those produced by archipelagic states) simply marking low water lines which are then omitted from other charts. Prescott (1987) has already indicated the degree to which Article 7 (LOSC) has been abused by state practice, and sea level rise could exacerbate this divisive tendency.

The merits of an interventionist approach by the parties to the Climate Convention should be apparent. Maritime boundary disputes are already a major feature of international relations and have generated a virtually unprecedented rise in submissions to settlement by the International Court and by Tribunals. At the time of the negotiation of the Law of the Sea Treaty regime there was no widespread recognition of the possible problems of sea level rise. The existing rules, while permitting flexibility, are not designed to accommodate the sort of radical physical changes discussed above. Formal acknowledgement of the potential problems of loss of maritime zones, outlined above, in a sea level rise Protocol, together with the promulgation of a new rule, such as that suggested by Soons and discussed above, would be of considerable importance in the maintenance of orderly international relations at a time of stress.

TOTAL INUNDATION OF STATES

The complete or major inundation of low-lying states, particularly island states, also poses a number of novel legal problems. As discussed above, a number of states have been identified which could be rendered uninhabitable by a one-metre rise in sea level with the consequent increase in storm flooding and loss of fresh water 'lens'. The question is whether such a state (such as the Maldives) could continue to exist if its total land area is covered by the sea, or so much of it that it cannot sustain sufficient population to have 'an economic life of its own'. Lewis has pointed out the problems of building traditional sea defences for Fanafuti, the main island of the Pacific Ocean State of Tuvalu on which 2,700 of the country's total population of 8,500 live:

> The land form of Fanafuti is so narrow and attenuated that in order to protect its 2.5 square kilometres, 54 kilometres of sea defences would have to be constructed. Moreover, so narrow is the land form for much of its length that sea defences on one side would be protecting the back of sea defences on the other with nothing in between.
>
> (Lewis 1990)

In addition, the land behind the defence would have to be raised to prevent it being flooded from within due to the porosity of the coral rock. As reports on the Maldives and Kiribati indicate, the situation is similar in many other coral islands (Edwards 1989; McClean 1989).

If the only land territory of the state becomes uninhabitable then the existence of the state itself becomes questionable. The 1933 Montevideo Convention lays down criteria that are regarded as a statement of the requirements of general international law. These criteria suggest that a 'defined territory' is a vital component of statehood. However, there are *sui generis* international persons, such as the Royal Order of Malta which do not meet this or some of the other of the Montevideo criteria. Recognition by other international legal persons has been seen to be the most important factor. It is clearly possible for an inundated state to maintain its recognition by other states as an international person *sui generis*. If such a state were to maintain its juristic existence, then the same criteria should be applied to the maintenance of its right to exercise sovereignty or sovereign rights over the maritime areas previously recognised as belonging to it, as would apply to the disappearance of baseline points discussed above.

Space does not permit a consideration of the issue of whether there are any customary international law obligations on neighbouring, or indeed any, state to take in what have been called 'eco-refugees' – the refugees from states and territories inundated by rising sea level. It is however worth recalling that the main international instrument on this subject, the 1951 Convention on the Status of Refugees, defines refugees solely in terms of religious, political, ethnic or other forms of persecution. In order to provide territory for such 'eco-refugees', it would presumably be possible for discrete areas (e.g. offshore islands or coastal areas) of existing states to be allocated to such refugees to be run on an autonomous basis. Soons (1990) argues that the most satisfactory solution in such a case would be the merging of the state of the refugees with that of the host, so as to create a new state which might claim the right to jurisdiction over the maritime zones and resources of the inundated state.

It should however be borne in mind that the rise in sea level will be a gradual and an episodic process. The Commonwealth Secretariat Expert Group reported that their studies suggested 'a variety of ways in which ... it would be possible for many inhabited low lying islands to adapt, albeit with external assistance' (1989). Although the problems of complete inundation discussed above should not be ignored, the short- and medium-term requirements are for effective cooperative planning. Customary international law, as Soons (1990) suggests, does provide mechanisms to meet many of the novel situations presented by the effects of sea level rise, but the present authors believe that planned adaptation will take place most effectively within the context of a new treaty framework.

COASTAL ADJUSTMENTS TO SEA LEVEL CHANGE

The latter part of this paper considers briefly the implications of coastal adjustments under rising sea level. Consideration of the land/sea boundary constitutes a wider view of the concept of a boundary than is usually taken by legal writers, but such coastal adjustments can have important impacts on lateral political boundaries as well as having other international law implications. These have been considered by the present writers in more detail elsewhere (Freestone and Pethick 1990; Freestone 1991).

It is clear that an increase in water temperature and mass will not simply result in the gradual flooding of low-lying land areas; indeed, in most regions the predicted rise in sea level will not be significantly different from the extreme levels presently experienced during storms. The change in mean sea level will, however, cause an increase in the periodicity of such extreme events and this will, in itself, create major problems for coastal areas. Most coastlines have, over the past 6,000 years since our present sea level was established, become adjusted to the periodicity of extreme events. Thus, a very high water level caused by storm waves may result in erosion of beaches, dunes or coastal wetlands – but these features will become re-established by the natural coastal processes in the relatively long period of time between such rare events. If the incidence of such high water levels increases, due to sea level rise, the reduction in the time available for coastal recovery between such events will mean that the present coastline suffers progressive erosion. Under a completely natural regime, such erosion would not continue indefinitely but would result in the establishment of a new equilibrium coastline whose landward displacement will mean that sufficient time is available for recovery between high water levels.

There are several problems associated with such an idealised development. The first is that natural processes, having been profoundly altered by human intervention, do not apply over most of the world's coasts. The most important of these alterations is the reduction in sediment supply to the coast, both by the prevention of coastal erosion and the construction of dams and reservoirs on inland rivers. Without such free movement of sediment, in many cases across local, regional and national boundaries, the coast will be unable to respond to the imposed changes in sea level. Second, it will be difficult to persuade the inhabitants of coastal areas that coastal erosion will ultimately lead to a restoration of stability and, moreover, if sea level continued to rise due, for example, to the lack of controls on the emission of 'greenhouse' gases, then the coastline will, indeed, be unable to attain a new equilibrium but will continue to erode landwards. Here again, international cooperation is required if the coastline is to achieve eventual stability. Finally, the adjustment of the coastline to a changed water level regime will not be confined to landward displacements; changes in the pattern of wave approach resulting from an increase in water depth will mean that the plan-view of the coastline will alter and, as discussed below, such changes may have important consequences for adjoining coastal landowners, regions, or even states.

These changes must be taken into account in any measured response to the problem of rising sea level. It is important for national authorities when developing their policy responses to such rises, to appreciate that the decisions that they take should not be taken in isolation. In areas such as the North Sea or East or West Africa where a number of states lie opposite or adjacent to each other with coastal frontages on a single sea, the strategy adopted by one state may have a considerable impact on neighbouring states. Indeed, some policy options may even exacerbate the problems of neighbours.

In its 1990 Report to IPCC Working Group III, the Coastal Zone Management Sub Group suggested that:

> The responses required to protect human life and property fall broadly into three categories: retreat, accommodation and protection.
> *Retreat* involves no effort to protect the land from the sea, the coastal zone is simply abandoned and ecosystems shift landward ...
> *Accommodation* implies that people continue to use the land at risk, but do not attempt to prevent the land from being flooded ...
> *Protection* involves hard structures such as sea walls and dikes, as well as soft structures such as dunes and vegetation, to protect the land from the sea so that existing land uses can continue.
>
> (IPCC 1990)

The two major contrasting policies, those of Protect and Retreat, provide a useful framework for analysis of some general legal implications. Crudely put, these two responses may be seen as *active* and *passive* policy responses, the implications of each of which will be examined briefly.

PROTECTION

Current thinking among coastal geomorphologists is that sea wall, embankment or bulkhead building is not the most effective form of response to erosion. The replacement of a wide, natural coastal zone over which wave energy is dissipated, by a linear wave defence such as a sea wall, inevitably leads to the ultimate destruction of the defence works and, perhaps more important, to the prevention of natural adjustments of the coastline to environmental changes such as sea level rise. The high cost of maintaining such hard defences once they have been deployed is, therefore, difficult to avoid, since their removal would result in rapid and major changes in coastline configuration.

The effect of such hard defences is, however, not confined to the coastal zone they were designed to protect. Instead, the cessation of natural erosional processes in one area of coast means that adjoining coastal areas will be deprived of their sediment supply. In areas of coastal erosion this can lead to an acceleration of the natural erosion rate: for example, in one well-documented case on the east coast of England the erosion rate along 15 km of unprotected coast adjoining a section of sea wall increased by four times over a 50-year period. In areas of coastal deposition such as estuaries, coastal wetlands and sand dunes, this deprivation of sediment caused by a bulkhead policy may reduce or prevent accretion which, in a period of sea level rise, may result in increased flood damage to low-lying coastal hinterlands (Pethick 1989).

It could be argued that, by analogy with river situation cases, action by one national authority in building a bulkhead which deprives another area in another state of sediment could give rise to liability under international law. In English law, such an action against a public authority under public nuisance would normally be covered by the statutory defence. However, in *Tate and Lyle v GLC and Port of London Authority* ([1983] 2 AC 509) the House of Lords said, in finding the defendants liable in public nuisance for building a ferry terminal that caused siltation of the plaintiff's jetty, that statutory operations must in general be conducted 'with all reasonable regard and care for the interests of other persons' (Paulden 1986).

At an international level, unilateral action by one state may have considerable effects on its neighbours. There are a number of examples. Sediment transport studies within the North Sea (e.g. McCave 1973; Eisma 1981; Dronkers, Van Alphen and Borst 1990) suggest that fine-grained sediment, derived from the cliff erosion on the east coast of

England, comprises an important source of suspended material in the North Sea which is vital for the accretion of coastal mudflats and marshes bordering Wadden Sea States. Here, a bulkhead policy by one state, which seeks to preserve the current coastal morphology, may seriously exacerbate the problems faced by its neighbours. Similarly, in West Africa, Shannon (1990) demonstrated that a combination of sea wall construction, harbour building and damming of inland rivers, by depriving the coast of sediment, has led to accelerated erosion in neighbouring states.

There are clear rules governing behaviour by one state which causes environmental harm to the territory of another, as for example in the *Trail Smelter Arbitration* (1938/1941) 3 (RIAA 1905). As Boyle (1991) points out, the *Trail Smelter* and the *Corfu Channel* (Merits) case ([1945] ICJ Rep 4) 'have long supported the proposition that no state may cause or permit its territory to be used to inflict serious harm on other States'. It seems clear, therefore, even in the current state of awareness of the global nature of the world ecosystem, that a state does not have the right to behave in a way which damages its neighbours even in its own territory. Most international environmental lawyers now agree that the behaviour of one state which adversely affects shared areas or even damages its own territory in a manner which affects the global ecosystem, gives rise to responsibility by that state to others, although, of course, a chain of causation must be proved to establish responsibility and indirect damage may be virtually impossible to prove.

Faced with such increasing criticisms against hard engineering structures on the coast, engineers and coastal managers have been made aware of recent developments in coastal protection which centre on the efficiency of the natural coastline in absorbing wave energy and thus preventing flood and erosion. For example, a recent UK Government report commenting on this approach states that:

> Increasing realisation of the effectiveness [of natural coastal systems] has led engineers to question the value of traditional coastal defences or protective structures. Instead they are examining solutions to coastal problems which seek either to control or emulate natural systems rather than replace them. This has become known as soft engineering – the solution of coastal management problems through environmentally sensitive schemes which involve substantial or total utilisation of natural systems as an integral part of the solution. These are more consistent with

ecological principles and other conservation interests as well as having greater social acceptability.

(Ministry of Agriculture, Fisheries and Food 1992)

Such an approach is not to be confused with the passive response to sea level rise advocated by the IPCC and discussed more fully below. The soft engineering approach to coastal protection attempts to simulate natural processes and to fine tune these towards some predetermined goal. This goal should both satisfy the requirements of the coastal user and achieve a long-term equilibrium form for the coast which requires no further management inputs. For example, since beaches are seen to be one of the most efficient ways of dissipating wave energy and achieving an equilibrium coastal form, a soft engineering approach might be to manipulate sediment transport processes to ensure an adequate supply of beach material at the required site. This type of approach depends on an understanding of the natural processes and their development towards an equilibrium coastal morphology and the existence of a clear goal for the project from the economic, social or conservation viewpoint.

It could be argued that the implementation of such soft engineering schemes, which provide coastal protection using natural systems, avoid many of the problems outlined above facing neighbouring coastal regions or states. It may be that, even in cases where their implementation does result in adverse interference with neighbouring coastlines, the fact that such changes can be regarded as a natural response to the environment would obviate any accusations of damage or responsibility.

RETREAT

A passive response is one which, at first sight, would appear to coincide with much of the current thinking of coastal geomorphologists, namely that coastlines will find a natural equilibrium when left to evolve naturally. There are two, quite distinct, roles for such a passive response to sea level rise. First, in coastal areas presently unencumbered by human infrastructures such as sea-defence works, reclaimed agricultural land or centres of population, it would be ecological and economic folly to interfere with the natural course of coastal evolution and here the passive response appears the only sensible course to take.

In areas where human infrastructure is present, a passive approach may not be taken so easily. In most of the populated coastal areas of the world, coastal reclamation has resulted in the human use of coastal areas which would, under a natural regime, act as a wave-energy absorbing zone. In order to utilise such areas, a variety of wave- and flood-defence structures have been provided. Removal of such defences or, under a totally 'passive' approach, allowing them to fall into disrepair, could only proceed after a careful cost-benefit analysis had demonstrated the value of these low-lying lands against the open-ended cost of defending them. On any cost-benefit analysis, low-lying cities would have to be defended: it would be economically – and politically – impossible to abandon London or New Orleans or the whole of the Netherlands.

In areas where cost-benefit considerations indicate that the cost of continued sea defence in a period of sea level rise is not a viable option, then coastal set-back or, to use a more positive phrase, coastal regeneration could be considered. This would, however, be a far from passive option. Reclaimed coastal areas, cut off from their source of marine sediment over many years, are in most cases considerably lower in surface elevation than the active coastal zone seawards of the sea wall, a factor exacerbated by land shrinkage due to drying and a variety of chemical changes. Abandoning the sea defences would thus create an area of open water which fails to absorb wave energy and which would result in accelerated erosion of the inner shoreline. Artificial recharge of these areas using, for example, dredge spoil, or careful management to allow natural accretion to take place, may provide a solution to these problems but it is far from being a passive solution.

In some populated areas, human use of the coastal zone has not resulted in major changes in the natural systems but rather has tended to utilise such systems for fishing, tourism, or recreation. In such areas a passive approach, which allows the natural systems to adjust to sea level rise without interference, may require considerable infrastructure displacement but this may appear, on cost-benefit grounds, to be the best option. The case of the North Norfolk coast in eastern England may be cited here as a good example of the difficulties facing such an approach. This coast is still largely in its natural condition; salt marshes, sand dunes and sand and shingle ridges exist here without restraining walls or groynes, since there has been relatively little development requiring such protection. Consequently, the area includes several SSSIs (Sites of Special Scientific Interest) and both local and National Nature Reserves, and has recently been designated as an International Wetlands Convention (Ramsar) site. The Ramsar designation for the area states: 'The primary management objective is to allow the natural processes of physiographic evolution and vegetative succession to proceed with minimal interference' (IUCN 1987). This management objective appears to be a call for the adoption of a passive approach as defined above. The history of the development of this coastline, including its development through several periods of sea level change since the last glacial period, offers no reason why this passive approach should not continue in the future, despite the predicted environmental changes. However, research into the changes in wave approach angles at this coastline, resulting from an increase in sea level, shows that the present pattern of salt marshes and beaches will not merely move landwards but will also migrate along the shore as the wave foci move with the new wave approach angles. Thus, areas now characterised by sandy beaches, which have generated a considerable recreational industry, will become coastal wetlands, while internationally important wildlife reserves, characterised by salt marshes and sand dunes, will revert to open sand beaches. Whether such changes will be tolerated either by the local or by the international community remains to be seen. A truly passive approach would necessitate the abandoning of the Ramsar wetland sites in the hope that areas of equal ecological importance would develop elsewhere but may, paradoxically, be seen in the short term as an abnegation of the Ramsar requirements.

Similar, although larger-scale, changes may take place on the Wadden Sea coastline of the Netherlands and Germany if a passive response to sea level rise is adopted. Here, changes in wave approach angles resulting from the rise in sea level could set up coastal currents which would accelerate the erosion of the more westerly Fresian Islands and transport the sediment eastwards into the German estuaries where deposition would occur. Attempts by the Dutch to preserve their important wildlife and recreational areas in the Fresian Islands would deny sediment to the German wetlands and thus prevent the adoption of a purely passive approach to sea level rise there.

Thus, the actions of one state which took a radically different approach or refused to cooperate with its neighbours in a common strategy could exacerbate the likely damage to its neighbour's coast. So, although a passive policy appears an attractive option, in most coastal areas it would not be possible to avoid continued intervention in coastal management. Moreover, the passive option could not be seen as a politically passive option. It would require considerable cooperation at a local and regional level. It would also have important impacts on, and legal implications for, the main coastal transitional zone wetlands. The implications of the two policies for wetlands have been considered elsewhere.

CONCLUSIONS

It should be clear from the inevitably brief discussion above that the range of issues to be addressed in the face of sea level rise requires cooperation. The present writers have argued elsewhere that this gives rise to an obligation among coastal states to cooperate which is not only a moral duty, but can also be argued to be a legal duty. It should also be clear that although existing international law does provide certain pointers to the future development of principles which will be necessary for an orderly and systematic response to sea level rise, the development of new rules and principles would be most effectively done within the context of a specific treaty regime.

There is, of course, always the possibility that predictions of climate change and consequently of sea level have been overestimated or are simply wrong. It is true that scientists are neither unanimous nor able to prove beyond doubt that it will take place. Decision-making in the face of uncertainty is always difficult, but given the likely scale and implications of the problem as predicted, a policy response based on the possibility that global warming will not take place does not present itself as unduly prudent. Indeed, the 1992 UN Framework Convention on Climate Change requires an alternative approach in its Article 3 which provides that 'where there are threats of serious or irreversible damage, lack of full scientific certainty should not be used as a reason for postponing ... measures.'

This is a commitment to what is known as the 'precautionary approach', an approach now adopted in an overwhelming majority of international forums concerned with the environment (Freestone 1991). The merit of its application in this context should be clear. By the time the evidence for global climate change is clear the world will be committed to changes which, even if they are reversible at all, will be far more expensive to arrest than measures taken at present. Existing customary international law does provide a basic structure of concepts and rules which can accommodate change, but the scale of projected changes and the level of sustained international cooperation which will need to be established suggests that treaty law would be a more appropriate vehicle. The Climate Change Convention concluded at the 1992 Rio de Janeiro Conference on Environment and Development is a framework convention for which specialised functional protocols will have to be negotiated. A Sea-Level Rise or Coastal Management protocol, as proposed by the IPCC Coastal Zone Management Subgroup (IPCC/CZM1990) may not, as yet, be a high priority but it is important that some form of cooperative framework be established before the impacts of sea level rise begin to manifest themselves and the needs of individual states become more immediate. In any event, this would not be a wasted effort. The level of cooperative effort in coastal conservation and management which a sea level rise protocol would envisage is surely a worthwhile enterprise in its own right.

NOTES

1 This chapter is a development of previous work, published as 'International legal implications of coastal adjustments under sea level rise: active or passive policy responses?' In J.G. Titus *et al.* (eds) *Changing Climate and the Coast* (Report to the IPCC from the Miami Conference, 1990), vol. 1, pp. 237–56. See also D. Freestone, 'International law and sea level rise' in Churchill and Freestone (1991). We are grateful to Graham and Trotman for their permission to draw on that chapter.

REFERENCES

Bird, E. and Prescott, J.R.V. (1989) 'Rising global sea levels and national maritime claims', *Marine Policy Reports*, 177:177–96.

Boyle, A. (1991) 'International Law and the protection of the global atmosphere: concepts, categories and principles', in R.R. Churchill and D. Freestone, *International Law and Global Climate Change*, London: Graham and Trotman/Martinus Nijhoff.

Churchill, R.R. and Freestone, D. (eds) (1991) *International Law and Global Climate Change*, London: Graham & Trotman/Martinus Nijhoff.

Dronkers, J., Van Alphen, J.S.L.J. and Borst, J.C. (1990) 'Suspended sediment transport processes in the southern North Sea', in R.T.Cheng (ed.), *Residual Currents and Longshore Transport*, New York: Springer-Verlag.

Edwards, A.J. (1989) *The Implications of Sea Level Rise for the Republic of the Maldives*, Report to the Commonwealth Secretariat.

Eisma, D. (1981) *Supply and Deposition of Suspended Matter in the North Sea*, Special Publications International Association of Sedimentologists, vol. 5: 415–28.

Freestone, D. (1990) 'Maritime boundary delimitation in the eastern Caribbean', *IBRU Conference Proceedings*, September 14–17, 1989, Durham, UK, 195–209.

—— (1991a) 'International law and sea level rise', in R.R. Churchill and D. Freestone (eds), op. cit.

—— (1991b) 'The Precautionary Principle', in R.R. Churchill and D. Freestone (eds), op. cit.

Freestone, D. and Pethick, J. (1990) 'International legal implications of coastal adjustments under sea level rise: active or passive policy responses?', in J. Titus and N. Psuty (eds), *Adaptive Options to Sea Level Rise*, IPCC/WMO/ UNEP.

IPCC/WGI (1990) Working Group I: Report.

IPCC/CZM (1990) *Strategies for Adaption to Sea Level Rise*, Report of the Coastal Zone Management Subgroup of the IPCC Response Strategies Working Group.

IUCN (1987) *Directory of Wetlands of International Importance*, Gland, Switzerland and Cambridge: International Union for Conservation of Nature and Natural Resources.

Kwiatkowska, B. and Soons, A.H.A. (1990) 'The entitlement to maritime areas of rocks which cannot sustain human habitation or economic life of their own', *Netherlands Yearbook of International Law*, 21.

Lewis, J. (1990) 'Sea level rise: some implications for Tuvalu', *Ambio* 18: 58.

McCave, I.N. (1973) 'Mud in the North Sea', in E.D. Goldberg (ed.), *North Sea Science*, Cambridge, Mass.: MIT Press, 75–100.

McClean, R.L. (1989) *Kiribati and Sea Level Rise, Report on a Field Visit for Commonwealth Expert Group on Climatic Change and Sea Level Rise*, London: Commonwealth Secretariat.

O'Connell, D.P. (1984) *The International Law of the Sea*, vol. ii, Oxford: Clarendon Press, 659.

Paulden, P. (1986) 'Ferry terminals as a public nuisance', *International Journal of Estuarine and Coastal Law*, 1: 70–4.

Pethick, J. (1989) 'Waves of change: coastal response to sea level rise', *Geographical Analysis*, 19: 1–4.

Prescott, J.R.V. (1987) 'Straight and archipelagic baselines', in G.H. Blake (ed.), *Maritime Boundaries and Ocean Resources*, London: Croom Helm.

Prescott, J.R.V. and Bird, E. (1990) *The Influence of Rising Sea Levels on Baselines from which National Claims are Measured*, Durham, UK: IBRU.

—— *Report of the Commonwealth Group of Experts: Climate Change: Meeting the Challenge* (1989), London: Commonwealth Secretariat.

Shannon, E.H. (1990) 'Coastal erosion and management along the coast of Liberia', in J. Titus (ed.), *Changing Climate and the Coast*, IPCC Working Group III: 25–48.

Soons, A.H.A. (1990) 'The effects of sea level rise on maritime limits and boundaries', *Netherlands International Law Review*, 37: 207–32.

Stoddart, D.R. and Pethick, J.S. (1984) 'Environmental hazard and coastal reclamation: problems and prospects in Bangladesh', in T. Bayliss-Smith and E.U. Wanmali (eds), *Understanding the Green Revolution*, Cambridge: Cambridge University Press.

7

UNITED STATES–RUSSIA MARITIME BOUNDARY

Robert W. Smith

INTRODUCTION[1]

On June 1, 1990, the Governments of the United States of America and the Union of Soviet Socialist Republics signed an agreement delimiting a maritime boundary.[2] By an exchange of notes on the same date, the two governments agreed to apply the provisions of the agreement, pending an exchange of instruments of ratification, effective June 15, 1990. On September 26, 1990, President Bush transmitted the agreement to the US Senate to receive its advice and consent to ratification (see Appendix B for full text of the agreement). On September 16, 1991, the US Senate voted favourably to give its advice and consent to ratification of the treaty.

With the break-up of the USSR in 1991 this agreement now pertains to the Government of Russia. As of May 1992, the Russian government had not taken further action on the agreement.

The boundary extends from the North Pacific Ocean through the Bering Sea and Bering Straits north through the Chukchi Sea and into the Arctic Ocean. The governments agreed that the 'western limit' as described in Article 1 of the US–Russia Convention of March 18/30, 1867 (hereinafter referred to as the Convention Line), in which the United States purchased Alaska, is the maritime boundary. Negotiations to agree on the proper depiction of the Convention Line began in 1981.

THE 1867 US–RUSSIA CONVENTION

The preliminary discussions that eventually led to the agreement by which Russia ceded Alaska to the United States were held in 1854 between Edouard de Stoeckl, Russia's newly appointed Chargé d'affaires in Washington (who in 1867 ultimately negotiated the cession Convention for Russia) and Secretary of State William Marcy and his friend, Senator William Gwin of California.

The Crimean War, begun in 1853 between Russia and Turkey, left Russia with serious financial problems, and in 1857 Grand Duke Constantine (adviser to his brother, Emperor Alexander II) suggested to Prince Gorchakov, the Foreign Minister, that Russia should sell to the United States the North American colonies (Alaska). But Russia did not make an offer at this time.

In 1860 Senator Gwin, who was interested in opening US trade in China, approached Stoeckl to ask if Russia would be willing to cede Alaska for five million dollars. This interest, approved by President Buchanan, was expressed in great secrecy to avoid aggravating the ongoing free-slave quarrel in Congress; the acquisition of Alaska would have been interpreted as an attempt to extend free soil. Stoeckl's report of this discussion to his government in St Petersburg reopened the question of cession. But by the time Stoeckl was advised to pursue the matter President Buchanan's domestic popularity had vanished and there was no chance of favourable action by Congress on any Administrative proposal. Stoeckl's offer would have to wait six years because of the outbreak of the US Civil War. On July 21, 1861, the day after the Battle of Bull Run, Stoeckl advised his government to abandon the negotiations.

American commercial interests in Alaska continued to expand during the 1860s. Russia's refusal to open Alaskan ports to foreign vessels upset US west coast fishermen. Official (by the new Secretary of State William Seward) and private overtures were made to Stoeckl to create better commercial access to the Russian territories. For political reasons – the continuing rivalry with Great Britain and the desire to maintain good relations with the United States – and on account of domestic economic concerns, the Russian government in December 1866 decided to pursue again the question of ceding Alaska to the United States.

Instructions were sent to Stoeckl who arranged to meet secretly with Seward. When Seward brought up the matter of commercial (hunting and fishing) rights in Alaska, Stoeckl replied that Russia would never allow either; the only alternative was outright purchase. Seward indicated his interest, but stopped short of making an offer, since he had no authority to negotiate. From this point in time events occurred quickly and secretly. On March 14, 1867, Secretary Seward told Stoeckl that President Johnson was not opposed to the transaction, but he had yet to present it to the cabinet. The following day Seward met with the cabinet, without the President's participation, and from notes presumed to have been on the table at that meeting:

Mr Seward proposes that Russia cede and convey to US her possessions on the North American continent and the adjacent Aleutian islands, the line to be drawn through the center of Bhering [sic] Straits and include all the islands East of and including Attoo [sic].

After making a few minor comments on the draft text, the cabinet unanimously approved the proposal.

On March 23, 1867, Seward and Stoeckl agreed on the substance of the purchase and on March 25 Stoeckl cabled the proposed agreement to St. Petersburg, requesting power to sign. He explained that Seward was most anxious to get the treaty before Congress, which was scheduled to adjourn in two weeks. On Friday, March 29, a cable reached Stoeckl authorising him to sign. That evening Stoeckl delivered the good news personally at Seward's home and he offered to meet the Secretary at the State Department the next day. Seward insisted on moving ahead quickly and after assembling his staff, the treaty was signed at 4 a.m. Saturday March 30, 1867. The Senate gave its consent to ratification on April 9; the Russian government ratified the treaty on May 15, and instruments of ratification were exchanged on June 20, 1867. The final price paid by the United States for Alaska was 7.2 million dollars.

THE MARITIME BOUNDARY DESCRIBED IN THE 1867 CONVENTION

The key article in the 1867 Convention pertaining to the maritime boundary is Article 1 which states:

The western limit within which the territories and dominion conveyed, are contained, passes through a point in Behring's straits on the parallel of sixty-five degrees thirty minutes north latitude, at its intersection by the meridian which passes midway between the islands of Krusenstern, or Ignalook, and the island of Ratmanoff, or Noonarbook, and proceeds due north without limitation, into the same Frozen Ocean. The same western limit, beginning at the same initial point, proceeds thence in a course nearly southwest, through Behring's straits and Behring's sea, so as to pass midway between the northwest point of the island of St Lawrence and the southeast point of Cape Choukotski, to the meridian of one hundred and seventy-two west longitude; thence, from the intersection of that meridian, in a southwesterly direction, so as to pass midway between the island of Attou and the Copper island of the Kormandorski couplet or group, in the North Pacific Ocean, to the meridian of one hundred and ninety-three degrees west longitude, so as to include in the territory conveyed the whole of the Aleutian islands east of that meridian.

Table 7.1 provides modern versions of several terms used in the above Article. Unfortunately, there was no official chart or map attached to the treaty depicting the exact course of the above definition.

EVENTS LEADING TO NEW NEGOTIATIONS

In March 1977 both the United States and the Soviet Union implemented 200-mile exclusive fishery zones.[1] Since the two countries' coasts are less than 400 miles apart, a portion of the North Pacific Ocean, Bering Sea and Chukchi Sea is overlapped by their respective 200-mile claims. Prior to March 1977 the two governments exchanged diplomatic notes indicating that each side would use the line established by the 1867 Convention as the limit for its extended fisheries jurisdiction.

THE NEGOTIATIONS

Eleven formal rounds of negotiations occurred between the two governments prior to the June 1, 1990 signing. Table 7.2 lists the dates and places of these rounds.

At the first round of talks it became evident that the two sides depicted the line described in the 1867 Convention differently. The

Table 7.1 Terms used in Article 1 of the 1867 Convention

Treaty term	Modern name
Krusenstern or Ignalook	Little Diomede Island (US)
Ratmanoff or Noonarbook	Big Diomede Island or Ostrov Ratmanov (Russian)
Behring straits or sea	Bering straits or sea
Frozen Ocean	Arctic Ocean
Cape Choukotski	Mys (Cape) Chukotski
Attou Island	Attu Island (US)
Copper Island	Ostrov Mednyy (Russian)
193°W longitude	167°E longitude

Table 7.2 Dates and places of the US-Soviet maritime boundary negotiations

Round	Date	Place
1	November 18-20, 1981	Washington, DC
2	May 17-19, 1983	Moscow
3	January 30-February 3, 1984	Washington, DC
4	July 23-25, 1984	Moscow
5	October 20-November 1, 1985	Washington, DC
6	April 3-4, 1986	Moscow
7	October 8-10, 1986	Washington, DC
8	October 5-6, 1987	Moscow
9	April 26-27, 1988	Washington, DC
10	September 17-19, 1989	Washington, DC
11	January 15-17, 1990	Moscow
Signing	June 1, 1990	Washington, DC

disagreement created a wedged-shaped area of approximately 20,868 square nm (sq. nm = 71,577 sq. km) between Convention lines defined by each side (Figures 7.1 and 7.2).

Unfortunately, Article 1 of the 1867 Convention did not specify the type of line that was to be used as the western limit. It should be noted that many modern-day maritime boundary agreements cite only 'straight lines' to connect a set of geographical coordinates, without indicating the type of straight lines. A 'straight line' can mean one of several types of lines; among the most common are a great circle arc, loxodrome or rhumb line, and geodesic. A brief definition of each follows:[4]

Great circle arc: a circle on the surface of the earth, the plane of which passes through the center of the earth. All meridians of longitude and the equator are great circles.

Rhumb line: a line on the surface of the earth making the same angle with all meridians; a loxodromic curve spiralling toward the poles in a constant angle direction.

Geodesic: a line of shortest distance between any two points on any mathematically defined surface.

Because of their defined qualities these 'straight lines' will appear differently on a given chart projection. The rhumb line, prominently used by navigators, is a straight line only on a Mercator projection; it appears curved on all other projections. On a Mercator chart, on which parallels of latitude and meridians of longitude all intersect at right angles, a great circle arc, which is the shortest distance between two

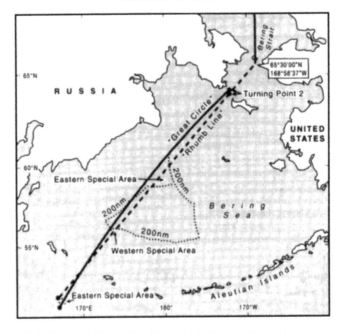

Figure 7.1 The 1867 Convention lines and the Russia–United States maritime boundary

points on a sphere, appears as a curved line unless it is a meridian of longitude or the equator. Great circle arcs and geodesics are very similar and the differences between the two, depending on the scale of the chart, may not be discernible.

The Soviet Union believed the lines defined in 1867 to be rhumb lines, while the United States believed that the negotiators meant the Convention line to comprise arcs of great circles. Given the high latitudes and the direction and distances of the Convention line segments the difference between the two types of lines is substantial. An area of almost 21,000 sq. nm is created by the overlap created by the two types of lines.

AGREEMENT IS REACHED

In January 1990 the two governments met in Moscow to finalise the agreement. General agreement was reached on the essential technical elements and a draft agreement was initialled. Further technical work was required following the meetings, and all work was completed for

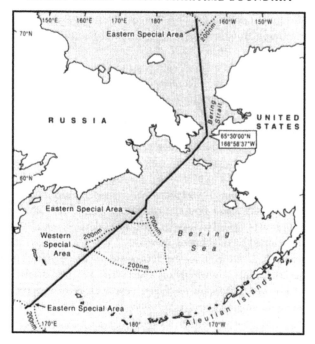

Figure 7.2 The 1990 United States–USSR maritime boundary

the June 1, 1990, signing by Secretary of State Baker and Foreign Minister Shevardnadze.

TECHNICAL CONSIDERATIONS

The parties had to address and agree on a series of technical questions. The first, and most fundamental, issue was on which spheroid and datum all calculations and depictions would be made. For the boundary technical exercise to proceed a common spheroid and datum was necessary, since references were to be made to coastal points on each side. Also, a common reference system was required to allow for an identical application of the final boundary.

The correct term for the figure of the earth is a geoid and to describe the geoid scientists turn to a *spheroid* (also known as an ellipsoid) which is a mathematically defined surface in which, among other parameters, the radius of the earth's major axis (a), the radius of the minor axis (b), and the degree of flattening (f), are known. To establish a 'best fit' of the earth's ellipsoid to the geoid for a given area of the world countries have developed different spheroids. The United States and Russia do not use the same spheroid for their respective national mapping and charting programs.

A geodetic datum is defined as a 'mathematical quantity or set of such quantities which serve as a reference or basis for other quantities'.[5] In geodesy two types of datum must be considered: the horizontal datum and the vertical datum. In a horizontal datum an origin would be established from which all other points, usually expressed in terms of latitude and longitude, could be determined.[6] A vertical datum establishes (i) a reference plane for elevations of land features above Mean High Water or Mean Sea Level and (ii) a lowest plane for referencing depths, i.e. tidal datum. Differences between neighbouring states of either datum could affect boundary delimitations. For this boundary delimitation, it was agreed to use the World Geodetic System (WGS 84), a system developed by the United States.

Under the WGS all the major geodetic systems can be unified and the coordinates of points anywhere on the earth are compatible. Using equipment known as Doppler receivers a user can receive signals from passing geodetic satellites and determine precisely the coordinates of unknown points relative to the WGS. Conversion equations make it possible to convert a given position back to a national datum. Ships carrying appropriate Doppler equipment can establish their position on WGS thus enabling an accurate determination of their position relative to the maritime boundary.

The two sides also agreed that for the final maritime boundary delimitation all 'straight lines', unless otherwise noted, would be geodetic lines.

THE MARITIME BOUNDARY

The description of the maritime boundary is found in the Annex to the agreement (see Appendix B). Except where noted in the agreement's Annex (i.e. the rhumb line segment 37–38 and the 200 nm arcs) the boundary turning points are to be connected by geodetic lines. Eighty-seven turning points were identified in order to aid any user in potting the maritime boundary regardless of the size or scale of a particular map or chart. The distance between any given pair of turning points is approximately 13–14 miles. The overall length of the boundary, from the Arctic Ocean to the North Pacific Ocean, is approximately 1,600 miles.

NOTES

1 The views contained in this paper are those of the author, and do not necessarily reflect those of the United States Government.
2 This boundary was negotiated with the government of the Soviet Union, but since the break up of the Soviet Union this boundary agreement will become a United States–Russian agreement. This paper was updated since the presentation in July 1991 to reflect the subsequent political events in Russia.
3 United States Law No. 94–265 (Fishery Conservation and Management Act of 1976), extending US fishing jurisdiction to 200 miles, effective March 1, 1977. The Supreme Soviet Decree of December 10, 1976, extending Soviet fishing jurisdiction to 200 miles also effective on March 1, 1977. Unless otherwise stated, all references to miles in this study are to nautical miles. One nautical mile equals 1.156 statute miles or 1,852 metres.
4 Defense Mapping Agency Topographic Center, *Glossary of Mapping, Charting and Geodetic Terms*, 3rd edition (Washington, DC 1973). For a more detailed discussion on this and other technical issues associated with maritime boundary delimitation, see Robert W. Smith, 'A geographical primer to maritime boundary-making', *Ocean Development and International Law Journal* (1982), 12(1/2), p. 1–22; Robert D. Hodgson and E. John Cooper, 'The technical delimitation of a modern equidistant boundary', *Ocean Development and International Law Journal* (1976), 3(4), pp. 361–88.
5 R.K. Burkard, *Geodesy for the Layman* (1968) (St Louis: Aeronautical Chart and Information Center), p. 10.
6 The origin of the North American 1927 datum, for example, which is tied to the Clarke 1866 spheroid, is located at Meades Ranch, Kansas.

APPENDIX B

Agreement between the United States of America and the Union of Soviet Socialist Republics on the maritime boundary

The United States of America and the Union of Soviet Socialist Republics (hereinafter 'the Parties'),

Recalling the US–Russia Convention of March 18/30, 1867 (hereinafter 'the 1867 Convention'),

Desiring to resolve issues concerning the maritime boundary between the United States and the Soviet Union,

Desiring to ensure that coastal state jurisdiction is exercised in all maritime areas in which such jurisdiction could be exercised for any purpose by either of the Parties, in accordance with international law, in the absence of a maritime boundary.

Have agreed as follows:

ARTICLE 1

1 The Parties agree that the line described as the 'western limit' in Article 1 of the 1867 Convention, as defined in Article 2 of this Agreement, is the maritime boundary between the United States and the Soviet Union.

2 Each Party shall respect the maritime boundary as limiting the extent of its coastal state jurisdiction otherwise permitted by international law for any purpose.

ARTICLE 2

1 From the initial point, 65° 30′ N, 168° 58′ 37 W, the maritime boundary extends north along the 168° 58′ 37″ W meridian through the Bering Strait and Chukchi Sea into the Arctic Ocean as far as permitted under international law.

2 From the same initial point, the maritime boundary extends southwestward and is defined by lines connecting the geographic positions set forth in the Annex, which is an integral part of this Agreement.*

3 All geographic positions are defined in the World Geodetic System 1984 ('WGS 84') and, except where noted, are connected by geodetic lines.

100

ARTICLE 3

1 In any area east of the maritime boundary that lies within 200 nautical miles of the baseline from which the breadth of the territorial sea of the Soviet Union is measured but beyond 200 nautical miles of the baselines from which the breadth of the territorial sea of the United States is measured ('eastern special area'), the Soviet Union agrees that henceforth the United States may exercise the sovereign rights and jurisdiction that the Soviet Union would otherwise be entitled to exercise under international law in the absence of the agreement of the Parties on the maritime boundary.

2 In any area west of the maritime boundary that lies within 200 nautical miles of the baseline from which the breadth of the territorial sea of the United States is measured but beyond 200 nautical miles of the baselines from which the breadth of the territorial sea of the Soviet Union is measured ('western special area'), the United States agrees that henceforth the Soviet Union may exercise the sovereign rights and jurisdiction that the United States would otherwise be entitled to exercise under the international law in the absence of the agreement of the Parties on the maritime boundary.

3 To the extent that either Party exercises the sovereign rights and jurisdiction in the special area or areas on its side of the maritime boundary as provided for in this Article, such exercise of sovereign rights or jurisdiction derives from the agreement of the Parties and does not constitute an extension of its exclusive economic zone. To this end, each Party shall take the necessary steps to ensure that any exercise on its part of such rights or jurisdiction in the special area or areas on its side of the maritime boundary shall be so characterised in its relevant laws, regulations, and charts.

ARTICLE 4

The maritime boundary as defined in this Agreement shall not affect or prejudice in any manner either Party's position with respect to the rules of international law relating to the law of the sea, including those concerned with the exercise of sovereignty, sovereign rights or jurisdiction with respect to the waters or seabed and subsoil.

ARTICLE 5

For the purposes of this Agreement, 'coastal state jurisdiction' refers to the sovereignty, sovereign rights, or any other form of jurisdiction with respect to the waters or seabed and subsoil that may be exercised by a coastal state in accordance with the international law of the sea.

ARTICLE 6

Any dispute concerning the interpretation or application of this Agreement shall be resolved by negotiation or other peaceful means agreed by the Parties.

ARTICLE 7

This Agreement shall be subject to ratification and shall enter into force on the date of exchange of instruments of ratification.
IN WITNESS WHEREOF, the duly authorized representatives of the Parties have signed the present Agreement.
Done at Washington, this first day of June, 1990, in duplicate, in the English and Russian language, each text being equally authentic.

For the United States of America: For the Union of the Soviet
James W. Baker III   Socialist Republics:
                Eduard Shevardnadze

* *Note*: Annex 1, giving geographical coordinates and exchange of notes in June 1990, and Annex 2, giving exchange of notes in January/February 1977 about fishing jurisdiction, are on p. 161.

8

MANAGING TRANSBOUNDARY FISH STOCKS

Lessons from the North Atlantic

Douglas Day

Prior to the establishment of exclusive economic zones (EEZs) in the late 1970s, most of the important commercial species were located in international waters (i.e. outside the jurisdiction of coastal states). The narrowness of coastal states' maritime zones meant that the amount of transboundary movement between adjacent states and between the high seas and territorial seas was limited. Some migratory fish moved from the high seas to territorial waters at certain seasons, as with cod off the coast of Newfoundland and Labrador and herring in the North Sea and Norwegian Sea: others crossed the boundary between territorial seas of adjacent coastal states, as in the case of herring between the Bay of Fundy and Maine. The most important transboundary (or straddling) stocks fell primarily under the management regime of the international community through regional organisations such as the International Commission for the Northwest Atlantic (ICNAF) or the Northeast Atlantic Fisheries Commission in European waters. The failure of these organisations to manage transboundary and other offshore stocks properly during the intense build-up of fishing effort in North Atlantic coastal areas prompted coastal states to declare 200 nautical mile (nm) exclusive economic/fishing zones in 1977 in an effort to safeguard the foundation of their inshore and offshore fisheries.

The creation of exclusive economic zones (EEZs) or exclusive fishing zones (EFZs) generally placed all the continental margin and, therefore, most of the world's important fish resources completely under the control of coastal states and thereby reduced or eliminated concern over the management of fish resources that crossed the former territorial seas/high seas boundaries. Although this specifically addressed most transboundary management issues arising from the overexploitation by distant water fleets in the 1960s and 1970s, the new EEZs and EFZs generated a wider range of transboundary management issues than had existed previously. In particular, they extended the length of maritime boundaries between adjacent and opposite coastal states and between coastal states and the high seas. In the process, many former high seas fish stocks were transformed into transboundary resources. The sheer length of the new jurisdictional boundaries[1] meant that the transboundary management problems had to be addressed on a much larger scale. In many regions of the world, the number of transboundary fish stocks increased as did the geographical extent of the area over which the problems had to be addressed. This focussed more attention on these stocks in regional fisheries management.

The UNCLOS III agreement provided general guidance on the management of transboundary resources by urging adjacent and opposite states to seek agreement upon 'measures necessary to co-ordinate and ensure the conservation and development of such stocks' and, in the case of stocks straddling the EEZ/high seas boundary, urging the coastal state and states fishing the area beyond and adjacent to its EEZ to agree on measures necessary to conserve stocks in the adjacent area (Article 63). Besides recommending that states with interests in these transboundary resources should seek to manage the resources cooperatively, UNCLOS indicates that there should be no discrimination in allocating quotas between interested states. By cross-referencing to other provisions of Part V of the agreement, it suggests that the proper conservation and management measures would not only ensure no overexploitation, but would also involve the determination of the allowable catch and the maintenance or rebuilding of each harvested stock to a level where they could produce the maximum sustainable yield (MSY) taking into account the needs of coastal fishing communities, fishing patterns, the interdependency of stocks, and other factors. Moreover, the coastal state is urged to promote the optimum use of fish resources in its exclusive zone, including giving access to other states where there are surpluses in the allowable catch (Article 62). In particular, when determining foreign access to any surplus, the coastal state is urged to consider the needs of developing states and the need to minimise dislocation in states that have traditionally fished in its zone.

Adherence to this package of international guidelines has varied from region to region and it is pertinent to review past achievements in transboundary stock management in order to suggest future courses of action. In this paper, four case studies with different regional frameworks are used to reveal the varied approaches to, and experiences in, transboundary stock management in the North Atlantic since 1977. These regional examples include the boundary between exclusive zones of opposite states in the case of the European Community and Norway, between adjacent countries in the case of Canada and the United States in the Gulf of Maine, and the coastal state/high seas boundary on the Grand Banks off Newfoundland. Reference will also be made, where appropriate, to the management problems created by the unresolved boundary dispute between Canada and France off southern Newfoundland.[2] It is not the intention to discuss the history of each of these cases in detail, as this has been done elsewhere,[3] but to use these cases as starting points from which to draw lessons for future management of straddling fish resources.

BACKGROUND

The European Community/Norway experience

As with the other cases under discussion, both the European Community and Norway established their EEZs in 1977. Although their claims overlapped, the common zonal boundary was automatically resolved with the application of the continental shelf boundary agreement reached in the mid-1960s. Cooperative management of the straddling fish stocks began immediately and a bilateral agreement was signed in 1980. Since then, bilateral negotiations on stock management for the following year have occurred annually. Scientific advice is sought jointly from the International Council for the Exploration of the Sea (ICES) as well as from each party's scientific advisers.[4] Each year total allowable catches (TACs) are negotiated for most stocks and shares allocated to Norway and the European Community. Early in the history of cooperative management, both parties agreed to allocate shares of demersal stock annual TACs according to the *zonal attachment* of the stock. More difficulty has been experienced with sharing the allowable catch of pelagic species, whose migratory habits often change over a period of years. It was not until the mid-1980s that agreement was reached over the allocation of herring shares according to a sliding scale related to the size of the spawning stock biomass (SSB). However, no agreement has been reached over the largest part of the transboundary mackerel resource (the Western Mackerel) which the European Community refuses to recognise as a joint stock. Until the 1980s, this stock had resided almost completely in European Community waters but subsequently the centre of gravity of fishing has shifted northwards with the changing stock migration pattern: now a large part is located in the Norwegian sector in summer months (Figure 8.1). The blue whiting has created other problems, for it is not limited in its transboundary migrations to the European Community and Norway and is the subject of multinational negotiations at the Northeast Atlantic Fisheries Commission.

Annual fish allocations to Norway and the European Community for most of the transboundary stocks form part of an overall agreement on fishing which involves reciprocal access, balancing the amount of fish taken by each party in the other's EEZ, and quota swapping. In accordance with the aims of the 1980 bilateral agreement, the parties allow specified reciprocal access to each other's zone in order to maintain historic fishing patterns. However, it is recognised that there must be an overall balance in the amount that Norway and the European Community take from each other's zone. Fishermen from the European Community are allowed to fish on the Norwegian side of the boundary for specified species, to maximum allowable amounts, in agreed geographical areas, and vice versa. When fishing in the other party's zone, the fishermen are subject to that party's technical regulations, enforcement, and monitoring.

The Gulf of Maine

The United States and Canada were involved in a jurisdictional dispute in the Gulf of Maine/Georges Bank area from 1964 to 1984. By the time both countries declared EFZs, their continental shelf dispute had not been resolved and their exclusive zone claims only served to accentuate the disagreements over the Gulf's resources. Two attempts to reach agreement on fisheries management for stocks straddling the disputed zone failed to be implemented[5] or approved[6] in the late 1970s, so that there was no coordinated management of these stocks before 1984 when the International Court of Justice decided the zonal boundary (the so-called 'Hague Line'). Since then, both countries have regarded the zonal boundary as a wall, allowing them to manage transboundary stocks according to their own management philosophy and without reference to the other's actions. These management philosophies differ markedly, as do monitoring and technical measures in each zone.[7] From 1984 to early 1991 there were many violations of the boundary by American fishermen, but now both countries have agreed on similar fines for these violations.

Canada/France

Management of stocks moving across the disputed area to the south and west of St Pierre and Miquelon (Figure 8.3) has been a source of considerable Canadian concern since 1984, when the trawler fleet from metropolitan France redeployed much of its effort from the Gulf of St Lawrence to the disputed area.[8] Although an interim fisheries agreement now controls French fishing in the area until the boundary is resolved,[9] the two countries do not agree on allowable catches, or the extent to which the stocks are being overfished.[10] However, they have agreed to exchange data on a regular basis about the amount caught by each country from stocks that will eventually straddle the new boundary. The management philosophies of both countries differ markedly and will doubtless be a source of disagreement when the new boundary is announced. Complicating the management picture are French treaty rights, currently controlled by a 1972 agreement which allows Newfoundland and St Pierre coastal vessels reciprocal access to each other's waters.[11]

Canada/high seas boundary problems on the Grand Banks

When Canada declared its 200 nm EFZ, it left portions of the continental shelf on the Grand Banks as international waters. Important groundfish stocks, including cod and American plaice, cross this outer limit of Canada's jurisdiction. International fishing effort has concentrated on these stocks at the Nose and Tail of the Bank, the adjacent areas to Canada's EFZ. As foreign access to the Canadian zone has become more restricted, the total fishing effort in these areas has increased and caused considerable concern for Canadian fisheries authorities. Management of the transboundary stocks is the joint responsibility of Canada and the Northwest Atlantic Fisheries Organisation (NAFO). There have been differences in their management philosophies, but of most concern has been the inability of NAFO to control both European Community fishing in the adjacent areas (any member of NAFO is allowed to opt out of the organisation's management decisions) and the actions of non-members.

PROBLEMS IN COOPERATIVE MANAGEMENT

The UNCLOS agreement urges states sharing transboundary resources to seek agreement on coordinated management for the conservation and development of these stocks (Article 63). The four cases considered suggest that three factors will play key roles in determining the extent of

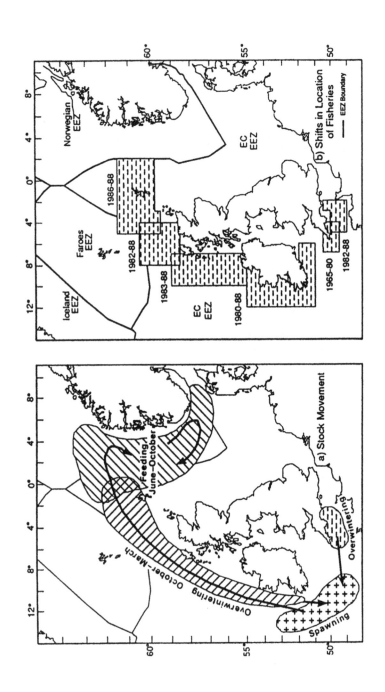

a) Stock Movement

Overwintering October–March

Feeding June–October

Overwintering

Spawning

b) Shifts in Location of Fisheries

— EEZ Boundary

Norwegian EEZ

Iceland EEZ

Faroes EEZ

EC EEZ

EC EEZ

1986-88

1982-88

1983-88

1980-88

1965-80

1982-88

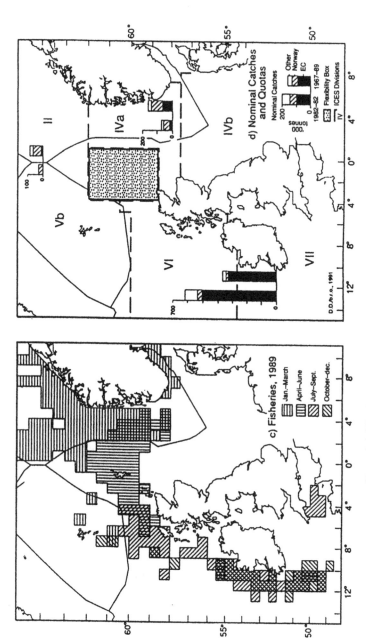

Figure 8.1 The Western Mackerel stock in EC–Norwegian waters

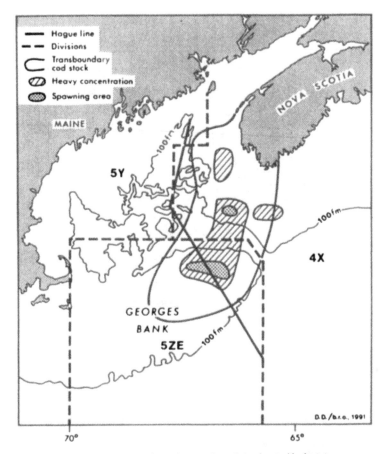

Figure 8.2 The transboundary cod stock in the Gulf of Maine

cooperative management: the history of the boundary delimitation, the desire to maintain historic fishing patterns, and the willingness to recognise and negotiate about transboundary stocks. These three factors are interrelated but, for the sake of simplicity, will be considered separately.

The nature of the boundary delimitation

The degree of cooperative management is conditioned by the attitude towards the boundary and its functions which, in turn, is related to the history of the boundary. The pre-existence of a continental shelf boundary between Norway and the European Community allowed quick

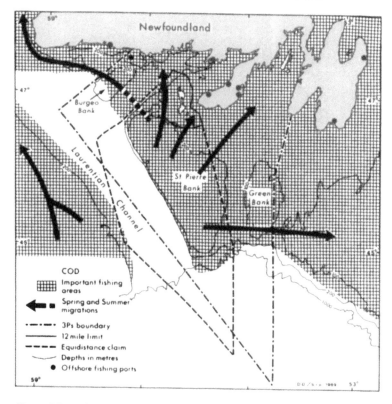

Figure 8.3 Cod migrations in the Newfoundland–St Pierre and Miquelon area

agreement on their common EEZ boundary so that attention could focus immediately on the management of their common fish resources and on finding a way to minimise the effect of the boundary on historic fishing patterns of each party's fishermen. On the other hand, the long-standing dispute over the Canada–US boundary in the Gulf of Maine and the eventual imposition of a boundary by the International Court in 1984 have made cooperative management of straddling stocks difficult, led to violations of the boundary since 1984 (principally by American fishermen who feel that the Hague Line was an unfair decision), and encouraged Canada and the USA to treat the boundary as a wall on each side of which was a zonal preserve controlled and managed entirely by the managers in one country without recourse to the initiatives of the other party.

The Gulf of Maine experience suggests that the Canada–France

111

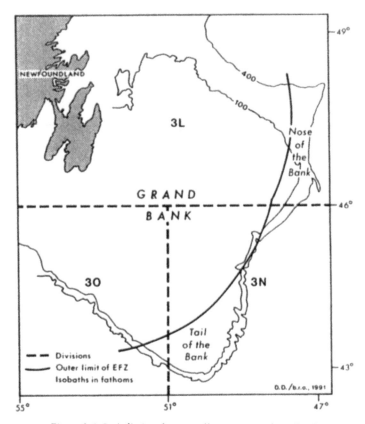

Figure 8.4 Jurisdictional zones off eastern Newfoundland

dispute around St Pierre and Miquelon, which has been longer and of late more bitter than the Gulf of Maine dispute, will set the framework for a prolonged and difficult period of transboundary stock management after the arbitration's boundary decision is handed down in 1992. How much cooperation there will be between France and Canada after 1992 will depend on the exact nature of the boundary drawn by the arbitrators, which country is most aggrieved by the result, and the impact on the 1972 agreement covering Canada–France fisheries relations. Uncoordinated management will probably be to the detriment of the cod fisheries of southern Newfoundland inshore waters, the southern Grand Banks, the northeastern Gulf of St Lawrence, and the offshore fisheries around the French islands. Clearly, if coordinated transboundary management is to occur and to be successful, it is imperative to concentrate attention on a quick and amicable solution to boundary disputes.

The boundary on the Grand Banks has also conditioned the attitude to management of those stocks. In this case, many Canadians feel that the areas beyond 200 nm should really be under Canadian jurisdiction[12] and Canada has considered extending the EFZ limit to include the Nose and Tail (Figure 8.4). Certainly it feels that, if it has the capacity to catch all the allowable catch of a straddling stock inside the boundary, it should do so and the international community should refrain from catching that species outside the boundary. This is clear from its advocacy of a moratorium on fishing at the Nose of the Bank. On the other hand, the European Community feels that, as Canada has control of most of the stocks off its east coast and as western European fishermen are now excluded from traditionally important fishing areas inside Canada's EFZ, it is legitimate for them to fish more intensively on the Nose and Tail in order to accommodate their displaced fleets.

Historic fishing patterns

The degree to which cooperative management is sought is affected by the historical pattern of fishing. The waters off Norway's North Sea coast were major destinations for the fishing fleets from European Community member states long before the move to extend maritime limits to 200 nm.[13] Similarly, important sectors of the Norwegian fishing industry depended on resources in what are now European Community waters. The imposition of a zonal boundary without some form of reciprocal access arrangement to protect traditional fishing patterns would have caused major disruptions for both sets of fishermen and the processing/marketing sectors of their industry. The demand for reciprocal access arrangements necessitated some form of cooperative management. On the other hand, the east coast North American industry is more limited to its own waters. Although some American fishermen from the New England states did fish in what are now Canadian waters off Nova Scotia and Newfoundland, this was not a major part of the New England fishing industry. Moreover, very few fishermen from Canada fished off the US coast in waters that are now inside its EEZ. Hence the imperative for a reciprocal arrangement was not so great as in the case of the European Community and Norway. Given also the intense feelings over the Gulf of Maine boundary in both countries, it is not surprising that the 1977 Reciprocal Fishing Agreement failed to be implemented.

Reciprocity is a feature of the 1972 agreement between France and Canada that was threatened in 1978–9, as the boundary dispute around St Pierre and Miquelon became more bitter. A precondition for sending the dispute to binding arbitration was that France would agree to some form of coordinated management in the disputed boundary area south of St Pierre and Miquelon, only if its offshore fishing fleet were given access to other parts of the Canadian EFZ in order to offset the reduction in catch inside the disputed zone.

Recognition of transboundary stocks

Coordinated management of transboundary stocks assumes that the parties involved recognise that the stocks are straddling stocks and part of a single resource. This is not always the case. Inadequacy of scientific knowledge is not necessarily the problem, as the European Community/Norway situation with regard to the Western Mackerel illustrates. The European Community has been unwilling to recognise formally that the changed migration patterns of the adult Western Mackerel stock have made it a transboundary resource. It recognises this informally but, when it comes to annual negotiations with Norway over transboundary stocks, the Western Mackerel is excluded. Instead Norway notes the TAC that the European Community has set for Western Mackerel and then sets its own TAC. This is coordinated management at a primitive level: it is hardly the kind that will ensure proper management of fish stocks in most boundary situations.

SUCCESSFUL MANAGEMENT

Cooperative management may lead to successful management but this is not necessarily the case. Similarly, it is theoretically possible for successful management to occur without coordinated management. However, it is implicit in the UNCLOS agreement that successful management of transboundary stocks involves coordination in order to ensure the conservation and development of these stocks. How successful has management been in terms of the four case studies? Here there are three areas that need to be discussed: the goal of rebuilding transboundary stocks to a level that will sustain high yields or of maintaining them at such a level; the enforcement of agreed management decisions; the impact of technical measures on the stocks.

Rebuilding or maintaining stock levels

The ultimate objective of UNCLOS is to rebuild stocks to the level at which they will produce the maximum sustainable yield, and to maintain them at that level. ICES, together with a report prepared for a review of the European Community's Common Fisheries Policy in 1992,[14] indicate that 13 of the 14 major stocks in the Community area are overfished. These include the major straddling stocks along the European Community/Norway boundary. Moreover, the chairman of the European Community for fisheries indicated at the end of 1990 during the European Community/Norway negotiations for 1991 TACs that there would have to be massive cuts in the agreed TAC for North Sea cod in order to rebuild the stock. A 1991 report of the Massachusetts Planning Commission[15] indicated that most of the demersal and shellfish stocks on the American side of the Gulf of Maine were heavily depleted and that a drastic reduction of fishing effort is needed in order to rebuild those stocks. On the Canadian side, the allowable catches from the transboundary cod stock were reduced in the 1991 Atlantic Groundfish Management Plan in recognition of the fact that this stock is overfished. Canadian/NAFO attempts to manage the northern cod that inhabits the waters of the northern Grand Banks through the Labrador coast fall into two periods: before 1985 they agreed to rebuild the stock and adopted the 1976 ICNAF goal of 1.5–2.0 million tonnes for the size of the SSB by 1985. TACs and quota allocations were set within a framework of an $F_{0.1}$ management model, but the SSB was still only 0.5 million tonnes by 1985. In the period since 1985, Canada has advocated increasingly restrictive TACs and a moratorium on cod fishing at the Nose of the Bank on the basis that little of the northern cod resource lies in international waters, and all of the available TAC can be taken by Canadian fishermen. NAFO has acceded to a moratorium on fishing cod on the Nose of the Bank during the past five years, although the European Community has used the objection procedure to opt out of this NAFO decision.

These cases all indicate that fourteen years after most of these stocks became transboundary resources, and transboundary management began, no headway has been made towards rebuilding the stocks. Why? The answer seems to lie in the level of determination to rebuild the stocks by the parties involved in their management, and this in turn is related to the socio-economic pressures to maintain high TACs in the light of overcapacity within the fishing industry. This is reflected in the relationship between the scientific advice given for the management of transboundary stocks and the management decisions.

In the European Community/Norway and Canada/NAFO cases, scientific recommendations on TACs were initially aimed solely at rebuilding the stocks and used an $F_{0.1}$ or modified $F_{0.1}$ management strategy. However, these recommended TACs required a substantial reduction in fishing effort. At least some decision makers in both cases found it difficult to accept these recommendations in the light of social and economic pressures from the fishing industry and so, explicitly or implicitly, they have set aside the objective of stock rebuilding by setting TACs and allowable quotas above the TACs suggested by the scientific advisors. In Canada strong pressures have led politicians to allow catches well above the recommended TAC for the northern cod resource,[16] part of which moves across the EFZ boundary to the Nose of the Bank. In NAFO, TACs set at the $F_{0.1}$ level have involved major reductions in the allocations to individual members: for the European Community such major reductions are resisted because of the size and importance of the Spanish and Portuguese distant water fleets, and so the European Community has opted out of the NAFO management decision and set its own allowable catch for the Nose of the Bank and other groundfish fisheries beyond the EFZ boundary. In recent years there has been a move inside NAFO, led by the European Community, to have the scientific advisors recommend alternative management strategies based on recent catch levels (e.g. F_{1990}). These management options are designed to keep TACs at a level in keeping with fleet capacity rather than to rebuild the stocks or to maintain them at their current levels. Ultimately this management strategy will lead to a further decline in the resource from the already low stock levels.

A similar situation has prevailed in the European Community/Norway case. Here again, the early scientific advice was based on the $F_{0.1}$ rebuilding strategy but in the late 1980s the European Community asked the ICES scientific advisory committee and its own scientific and technical advisors for management options beyond this strategy. In the management of transboundary stocks, agreed TACs have normally been set above the recommended levels from scientific advisors primarily in response to pressure from within the member states for higher allocations for their oversized fleets.

Clearly the lesson is that, if the transboundary stocks are to be rebuilt in accordance with the international community's wish as expressed in the UNCLOS agreement, then there must be a greater determination on the part of decision makers to adhere to rebuilding strategies. This, in turn, makes it imperative that short-term socio-economic pressures must be resisted in order to achieve improvement of the stocks and the possibility of increased catches over the long term. For this to occur, there must obviously be a reduction of fishing capacity in all states fishing the transboundary stocks.

Setting allowable catches and national quotas

The UNCLOS agreement recommends that the parties involved in managing transboundary stocks establish allowable catches and implies that these should be distributed fairly. The four case studies show a mixture of circumstances with regard to the establishment and allocation of catches.

In the European Community/Norway case, agreed total allowable catches are set for most, but not all, transboundary species. As already noted, these TACs are normally set above the level at which the rebuilding strategy (e.g., $F_{0.1}$) would recommend and, in the case of Western Mackerel, the effective TAC is the total of the individual European Community and Norwegian TACs. The same situation regarding the effective TAC prevails in the St Pierre and Miquelon case where Canada sets an overall TAC (including an allocation for France) for the disputed zone but France does not recognise the Canadian allocation and so sets its own TAC. In the Canada/NAFO case, agreed TACs are set for each transboundary stock but the allowable catch is effectively increased by a member that opts out and sets its own allowable catch and by non-members of NAFO who are not bound by that organisation's management policy. Finally, there are situations where no effective TAC is set. This is the case in the Gulf of Maine, where only one party (Canada) subscribes to the view that an allowable catch is a desirable part of a management strategy. Canada sets allowable catches for the part of the stock on its side of the boundary, but the US management strategy does not require specification of an allowable catch. Clearly, if there is to be an allowable catch for a transboundary stock, it must include all parties and preferably should be set in a co-ordinated fashion to reflect the yield capability of a stock. Another major concern, however, is that if TACs are used they are usually set for individual species and without reference to the interrelationships between stocks. This is particularly important where TACs are set higher than recommended by scientific advisors, as raising the level of exploitation of one species may have further implications on other parts of the food web.

Allocation of that allowable catch between parties is another matter and has been the focus of different attempts at determining appropriate relative shares. There have been three basic approaches to this issue, two of them used in the European Community/Norway case. The idea of 'zonal attachment' was used to define the relative European Community/Norwegian shares of the five major demersal species (cod, haddock, whiting, saithe, and plaice). Scientists suggested the use of criteria such as the location of spawning, juvenile, and pre-recruit areas, the occurrence and migration of the fishable part of the stock, and the history of the fishery and present state of the stock's exploitation, to establish zonal attachment.[17] Once defined and agreed to by the managers, division of the annual TAC was simple and automatic. For stocks that are less stable in their zonal attachment, the method could still be used if the relative shares were reviewed after any major shift in zonal attachment. The European Community/Norway also uses a sliding scale for determining the relative shares for the division of the annual adult herring allowable catch. This sliding scale is based on the size of the spawning stock biomass (SBB).

An alternative method of determining relative shares of the allowable catch is used in the North American case studies. This is based on historical trends in the fisheries. In the case of the migratory stocks in the Canada–France disputed area, Canada set the French share at 15.6 per cent based on the traditional catches by the French fleet in this area. However, the French did not accept this allocation.[18] Under the 1979 Fisherie Agreement, the catch of groundfish and scallops for both US and Canadian fishermen in the Gulf of Maine was determined according to historic catches in the region. However, the US administration, under pressure from New England fishermen, refused to ratify the agreement primarily because the US scallop fishery had been expanding in recent years and its relative share in the agreement was set too low. Both cases illustrate the problem with using historic catches to set relative shares. The French fishery in the disputed area south of Newfoundland expanded after 1984 and so France would not accept a relative share based on pre-1984 catches. The same kind of problem was encountered by the Americans with regard to the 1979 Fisheries Agreement with Canada. It is difficult for countries to agree on the period of historical catches that reflects a normal situation when one country's industry is expanding. It could well be that the zonal attachment idea holds more potential for determining relative shares where scientific knowledge is adequate.

Balance

In situations where it is deemed desirable by the parties to have a reciprocal access agreement, in order that fishermen from a country may fish on either side of the boundary, it is often thought desirable to achieve a balance in terms of the fish taken from each other's zone. Kawasaki reveals that this was an important condition of the Soviet/Japanese agreement for the exploitation of the transboundary saury resource off northern Japan and the Kuril Islands.[19] The same has been the case with the European Community/Norway transboundary stocks. In this instance, balance is achieved by converting the fish taken from the other party's zone into a standardised unit (cod equivalency value) and then engaging in quota swapping in order to achieve the desired balance. An alternative to the European Community/Norway method of achieving a balance would be payments per unit of fish taken or per fishing day in the zone. However, none of the other cases reviewed have reciprocal access arrangements with the exception of that under the 1972 Canada/France fisheries agreement. In that case the quantities of fish involved are small and no attempt is made to balance the catches of Newfoundland and St Pierre boats in the other party's zone.

Probably the European Community/Norway model of achieving an overall balance is one that should be considered elsewhere but it would require modification of the standard unit of equivalency between regions and also should provide for periodic review of the standard unit of equivalency. In the European Community/Norway case there has been no revision of cod equivalency values since the standard was originally established in the late 1970s but, given the possibility that relative values of the major commercial species may change over time, it would appear desirable to allow for a revision of the value weightings over time.

MANAGEMENT UNITS

In all the cases studied, the areas used to formulate management strategies for transboundary stocks are statistical units used for the collection of catch/effort data rather than either the biogeographical areas of fish stocks or the boundaries between EEZs/EFZs. These statistical units straddle the zonal boundaries. In all cases, this can give rise to problems as the decision makers tend to manage fisheries resources by the statistical areas and do not necessarily see or recognise the connections between these management areas in their decision making. The exception is the Canada/US case, where Canada sought and obtained an adjustment of the boundary of one management area in the Georges Bank area (5Ze) in order to make the boundary conform with that of the Hague Line. However, in the European Community/Norway case the management areas in the North Sea divide the herring resource into several different areas. When the TAC for North Sea herring is set at the annual bilateral negotiations, it does not take into account, or apply to, the juvenile herring that inhabit the Skaggerak/Kattegat area and which are exploited in a Danish industrial fishery. Although the size of the herring spawning stock, which is used as a determinant of the TAC and the shares of the two parties, is greatly affected by the industrial fishery, the by-catch of herring in this fishery is not taken into account in the determination of the adult herring catch. In the long term, this situation could make it impossible to manage the North Sea herring successfully. It is suggested that, where possible, fisheries management areas should coincide with the biogeographical range of the major species and that, if this demands different management areas for each kind of fish, then new boundaries should be set.

Monitoring

Monitoring of transboundary fishing activities varies. A very high level of inspection by aerial surveillance and on-board inspection is carried out on a week-to-week basis by Canada in both the Grand Banks region and on its side of the Gulf of Maine boundary. A programme of on-board inspection by NAFO based on the Canada model is now being developed in the adjacent area of the Grand Banks. Although it is possible for fishing vessels to be missed at night or in fog or in between surveillance flights, new RADARSAT technology will soon provide economies in surveillance and extend this to night-time and foggy conditions.

Elsewhere, monitoring of the fishing activities on transboundary stocks is at a low level. On the US side of the Gulf of Maine, monitoring is carried out on board by some 13 officials between Georges Bank and Virginia. This is totally inadequate to enforce the fisheries management policy adopted in the Gulf of Maine. Within the European Community there is very poor monitoring by the European Commission as only national agencies have the capability to monitor fleet activities at this time. Although the European Commission negotiates the agreed annual TACs with Norway and allocates them between its member states, only the member states of the Community are able to monitor whether or not the allowable catches and technical measures (such as minimum landing sizes and mesh sizes) are being observed on the European Community side of the zonal boundary. It is known that misreporting of catches has reached a high level (e.g. in the mackerel box area and in the Skaggerak industrial fishery) and that illegal fish are landed (e.g. herring in the industrial catches in Denmark). As a result the European Community is proposing to develop its own monitoring system.

Technical problems across the boundary

Although the overall management of the transboundary resource may be subjected to bilateral negotiations, in accordance with the UNCLOS III agreement the day-to-day regulations are usually left to the country controlling each zone. This often leads to transboundary differences in the fishing techniques and gear size used, as in the case of groundfish straddling the European Community/Norway boundary. On the Norwegian side commercial species may be caught using 110 mm mesh, whereas on the European Community side 90 mm mesh is used. Differences in the configuration of the mesh (diamond versus square) may occur on either side of a boundary in a fishery involving the same straddling stock(s). Sometimes such a difference may be legitimate and acceptable in the management of a transboundary stock. This appears to be the case in the groundfish fisheries of the northern North Sea in the European Community zone and the Norwegian Trench area. In the latter area, the depth of water allows more specialised single species fishing than in the European Community waters. However, harmonisation would normally be desirable where there are no great environmental differences in terms of the water conditions and stock separation.

Enforcement

The four case studies suggest that a key element in the successful management of transboundary resources has to be a coordinated and effective programme of enforcement to ensure that TACs and other policy control measures are effective. Moves towards the harmonisation of enforcement policies in the Canada/NAFO and Canada/US cases recognise the importance of this aspect of transboundary stock management. The European Community/Norway case exemplifies the worst type of situation: enforcement is trying to achieve diametrically opposed goals.

Norway's policy forbids European Community vessels fishing in the Norwegian sector from discarding unwanted fish or by-catches (a regulation also applied to Norwegian vessels). However, European Community vessels returning to the European Community zone are not allowed to land undersized fish or unwanted species, but can discard these in European Community waters. Hence, it is possible for Danish fishermen fishing North Sea mackerel in the Norwegian zone to catch other fish, which they dump upon entering the European Community zone.

CONCLUSIONS

The UNCLOS guidelines and suggestions on the management of trans-boundary stocks imply that there should be *cooperative* management. At least three conditions determine the extent to which this will occur: the circumstances under which the boundary was established, the desire of each party to minimise disruptions to historic fishing patterns, and the willingness of each country involved to recognise and discuss management problems. These factors are interrelated. In areas where common EEZ boundaries are established without dispute and bitter-ness, and each country wants to minimise the disruption to historic fishing patterns, productive discussions on cooperative management can usually occur at an early stage. However, a prolonged and bitter dispute over the location of the boundary, as in the case of the Gulf of Maine, will probably make it difficult to maintain historic fishing patterns unless some pre-existing treaty guarantees continued reciprocal access, as is the case between Canada and France with respect to the coastal fishing vessels of both St Pierre and Newfoundland. In general, for cooperative management to be achieved in the short term, boundary disputes have to be settled quickly and with as little bitterness as possible.

If cooperative management is to minimise the effect of the boundary on historic fishing patterns, either some form of reciprocal access is required as in the case of opposite or adjacent states, or concessions must be made by the coastal state and the international community in the case of an EEZ/high seas boundary. In the case of a single individual transboundary resource where the parties involved want to maintain historic fishing patterns and where these involve fishing in the other party's EEZ/EFZ, the parties must agree on the relative shares that can be taken in total from the resource, regardless of where the fishing occurs in relation to the boundary.

Establishment of the relative shares for each state involved has caused problems in a number of instances. Three basic approaches have been used. Use of *historic catches* to determine relative shares between adjacent or opposite coastal states requires agreement on the time frame relevant to the decision. This is not always easy, especially when one or other state is expanding its fishery. Disagreement may undermine efforts at cooperative management, as in the case of the US–Canada 1979 fisheries agreement on the Gulf of Maine. The principle of assigning relative shares on the basis of *zonal attachment* holds much appeal for species whose migratory patterns remain stable over long periods of time and for others if provision is made for either periodic review of the degree of zonal attachment (for example, modelled on unitisation procedures applied to oil wells straddling boundaries in the North Sea) or review following a significant and observed change in the migration patterns. Periodic review would prove easier and be fairly automatic if a standard method of calculating zonal attachment (including a weighting for the various factors taken into consideration) were agreed at the outset. However, zonal attachment does demand a good scientific knowledge of the biogeographical characteristics of the stock in relation to the boundary. The third approach, assigning relative shares on a *sliding scale based on the size of the SSB*, is more complex and it is possible that one country's actions in juvenile fisheries may influence the share of the resource that can be harvested at the adult stage by another country.

UNCLOS III required the establishment of an allowable catch for each transboundary resource. Where cooperative management is lacking and where each state wishes to manage its part of the stock in isolation or in accordance with a biological model attaching little or no importance to the allowable catch, this may not happen. Under this set of circumstances, overfishing will tend to occur if there are no effort controls. The experience on the American side of the Gulf of Maine illustrates the point. It is perhaps worth noting that current moves by some groups of European Community fishermen to eliminate TACs/quotas may lead in this direction if vessel decommissioning is unsuccessful.

Cooperation is only one part of the framework for the successful management of transboundary stocks. In terms of the UNCLOS agreement, successful management would rebuild stocks to, or maintain stocks at, a level that will yield high catches over a continuing period. The UNCLOS agreement sets the target level as MSY. To achieve this goal, all parties must also be determined to follow a rebuilding or stock maintenance approach to management. This is unlikely to occur unless the preconditions of cooperative management exist, but these are only permissive and do not guarantee such an approach. Even in the case of only two parties with adjacent zones, this may not happen: experience shows that as the number of states involved increases (as in the case of the European Community or NAFO) the difficulties in adopting such an approach may increase.

This highlights a fundamental problem between scientists and decision makers. It is comparatively easy for scientists to recommend the required TAC for the rebuilding or maintenance of a resource: it appears to be universally difficult for decision makers to accept these recommendations in the light of social and economic pressures and so, explicitly or implicitly, they often set aside the objective of stock rebuilding. These pressures are strong within coastal, federal states such as Canada or the USA: in the context of a multinational organisation such as the European Community or an international organisation such as NAFO, the pressures may be much more intense. Nonetheless, the socio-economic pressures to allow higher catches must be resisted if successful management of transboundary stocks is to occur. This will be easier if the managers agree to reduce overcapacity. Even if they do accept and adopt these measures, each party must adopt enforcement procedures that are strong, effective, and harmonised, together with technical measures that are also harmonised. Ultimately whether or not the UNCLOS provisions for transboundary stock management are realised may depend not so much on what the full international community operating at the global level considers desirable, but on what is practically acceptable at the regional level given the similar political, economic, and social pressures on decision makers in widely differing regional situations.

ACKNOWLEDGEMENTS

Research for this paper was undertaken with a grant from the Saint Mary's University Senate Research Committee. The author would also like to thank the staff of NAFO, Christopher Southgate and other members of the UK Ministry of Agriculture, Food and Fisheries, Marius Hauge (Norwegian Embassy), members of the European Commission and the staff of Eurofish for their help during discussions and in supplying research documents.

NOTES

1 See, for example, map of jurisdictional zones in the Northeast Atlantic by H.D. Smith, 'Maritime Boundaries and the Emerging Regional Bases of

World Ocean Management' in G. Blake (ed.), *Maritime Boundaries and Ocean Resources*, Croom Helm, London, 1987, p. 83.

2 For further details see D. Day, 'St Pierre and Miquelon Maritime Boundary Case: Origin, Issues, Implications' in Carl Grundy-Warr (ed.), *International Boundaries and Boundary Conflict Resolution*, Durham: International Boundaries Research Unit, University of Durham, 1990, 151–74.

3 See (i) Oceans Institute of Canada, *Managing Fishery Resources Beyond 200 Miles: Canada's Options to Protect Northwest Atlantic Straddling Stocks*, Ottawa: Prepared for the Fisheries Council of Canada, January 1990, 87 pp.

(ii) J.L. Bubier and A. Rieser, 'US and Canadian Groundfish Management in the Gulf of Maine–Georges Bank Region', *Ocean Management*, 10 (1986), 83–124.

(iii) D. Day, 'North Atlantic Experience in Transboundary Stock Management' in J.L. Suarez de Vivero (ed.), *The Ocean Change: Management Patterns and the Environment*, Seville: Department of Human Geography, University of Seville and the International Geographical Union, 1992, 253–64.

4 The Commission of the European Communities has a Scientific and Technical Committee for Fisheries, which presents reports on a yearly or twice yearly basis.

5 The 1977 Interim Reciprocal Fisheries Agreement was not implemented because of disagreements over implementation procedures.

6 The 1979 Fisheries Agreement was not ratified by the United States' Congress.

7 See Bubier and Rieser, op. cit.

8 D. Day, 'The Saint Pierre and Miquelon dispute: towards a further re-definition of French fishing rights in the Northwest Atlantic', *Ocean and Coastal Management* 18 (1992), 371–403.

9 *Agreement between the Government of the Republic of France and the Government of Canada relating to fisheries for the years 1989–91*, Paris and Ottawa: Proces-Verbal, Annex II, March 30, 1989.

10 See (i) Government of France Note Verbale (France), Paris: March 30, 1989.

(ii) Department of External Affairs Canada Note No. 220, Ottawa: March 30, 1989.

11 *Agreement between Canada and France on their Mutual Fishing Relations*, Ottawa: March 27, 1972.

12 Oceans Institute of Canada, op. cit., 82–5.

13 See J. Coull, *The Fisheries of Europe*, London: Bell and Sons, 1972.

14 Commission of the European Communities, *Communication from the Commission to the Council and the European Parliament: Common Fisheries Policy*, Brussels: Sec (90) 2244 final, December 6, 1990.

15 'New England Groundfish Study: Need for quotas, limited entry, trip limits, etc. recommended', *The Sou'wester*, vol. 23(9), February 1991.

16 Northern Cod Review Panel, *Independent Review of the State of the Northern Cod*, Ottawa: Department of Fisheries and Oceans, February 1990.

17 ICES, *The Biology, Distribution, and State of Exploitation of Shared Stocks*

in the North Sea Area, Cooperative Research Report No. 74, Charlottenlund Slot, Denmark: 1978.

18 See (i) Government of France Note Verbale (France), Paris: March 30, 1989.
   (ii) Department of External Affairs Canada Note No. 220, Ottawa: March 30, 1989.

19 T. Kawasaki, 'The 200-mile Regime and the Management of the Trans-boundary and High Seas Stocks', *Ocean Management*, 9, 1984, 7–20.

9

THE OLD AND NEW EGYPTIAN LEGISLATION ON STRAIGHT BASELINES

Giampiero Francalanci and Tullio Scovazzi[1]

HISTORICAL OVERVIEW OF THE EGYPTIAN INTERNAL WATERS

It may perhaps be of some interest to consider the internal waters of Egypt in a historical perspective, before coming to recent developments.[2]

In a reply given in 1927 to a question from the Committee for the Codification of International Law of the League of Nations, Egypt considered the Gulf of El Arab, where the important city of Alexandria is located, as 'totally comprised within the territorial sea'.[3] A hypothetical closing line of the Gulf of El Arab, taking as its entrance points Ras el Daba and the western cape of the bay of Abu Qir, is about 94.7 nm in length and includes a point whose distance from the nearest point on the shore is about 25.7 nm.

The legal basis for the 1927 Egyptian reply was however questionable. El Arab was not regarded as lying within the Egyptian internal or historic waters, but rather as within its territorial sea.[4] No subsequent official legislation of Egypt clarified its status. In a 1951 note the United Kingdom expressed the view that no historic bay is situated in Egypt.[5]

As regards straight baselines, the Egyptian practice is derived from a Royal Decree of January 15, 1951,[6] which almost literally followed a Royal Decree of Saudi Arabia of May 28, 1949.[7] In the Egyptian decree there are significant provisions on bays and islands.

The term 'bay' includes any inlet, lagoon or other arm of the sea (Article 1b). The Egyptian baselines are 'where a bay confronts the open sea, lines are drawn from headland to headland across the mouth of the bay' (Article 6b). It is easy today to note that the Egyptian decree did not mention the two geographical conditions to be met in

127

order to qualify as a juridical bay under Article 7 of the Convention on the Territorial Sea and the Contiguous Zone (Geneva, April 29, 1958)[8] and under Article 10 of the United Nations Convention on the law of the sea (Montego Bay, December 10, 1982),[9] namely: the maximum length of 24 nm of the closing line and the semicircle rule, requiring that a bay must be at least as large as the semicircle whose diameter is a line drawn across its mouth.

Of course, it makes no sense to blame Egypt for not having in 1951 complied with rules that were included in a treaty of codification opened for signature in 1958. The relevant point is rather that, in 1951, things were moving as regards straight baselines. In 1935 Norway had challenged the restrictive attitude of the major maritime powers (first of all, the United Kingdom) by adopting the Royal Decree of July 12, 1935, which established a straight baseline system for the northern part of the country. In this region the coast is deeply indented and there is a fringe of islands, islets, rocks and low-tide elevations in the vicinity of the coast (the words *fjord* and *skjaergard* are self-explanatory). On December 18, 1951, the International Court of Justice rendered its judgment in the case brought by the United Kingdom against Norway and stated that the Norwegian straight baseline system was not contrary to international law.

The Egyptian attitude in 1951 may be seen as an attempt to propose straight baselines also in cases where – like the Egyptian Mediterranean coast – there is a succession of indentations of considerable breadth and relatively small depth. A commentary on the law of the sea published in Egypt in 1952[10] included, as examples, maps showing, besides the above-mentioned Gulf of El Arab, the Gulf of Salum (45.4 nm wide and 9.9 nm deep), Abu Hashaifa (31.6 nm wide and 7.9 nm deep), Pelusium (49.3 nm wide and 13.8 nm deep), and El Arish (65 nm wide and 11.8 nm deep) (Figure 9.1).

Such an attempt did not however prove to be successful. The 1951 judgment of the International Court of Justice was shaped on the geographical features of the Norwegian coast and served as a model for the wording of Article 4 of the Geneva Convention ('In localities where the coastline is deeply indented and cut into'). In its protest,[11] the United Kingdom objected to Egypt 'that a gulf must have a reasonable penetration inland in proportion to its width' and that it was prepared only to accept that the territorial sea may be measured from a straight line across a bay where the distance between the low-water marks on the opposite sides of the bay is not more than 10 nm. As already mentioned, Article 7 of the Geneva Convention introduced two precise

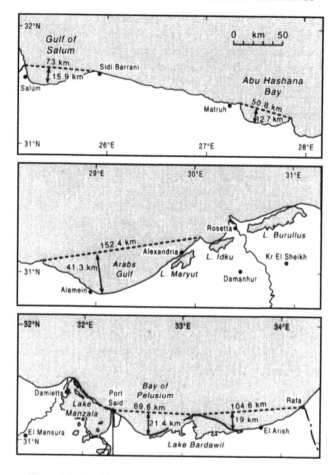

Figure 9.1 Possible application of the 1951 Decree – central
Mediterranean coast

geometrical conditions according to which an indentation can qualify as
a bay in a juridical sense.[12]

As regards islands and island-like features, the Egyptian Decree of
1951 includes in the terms 'island' 'any islet, reef, rock, bar or permanent
artificial structure not submerged at lowest low tide' (Article 1c) and
defines a 'shoal' as 'an area covered by shallow water, a part of which is
not submerged at lowest low tide' (Article 1d). It is clear that the
Egyptian definition of an island differs from the generally accepted
definition of an island as a 'naturally-formed area of land, which is

129

above water at high-tide' (Article 10, Paragraph 1 of the Geneva Convention). Moreover, the Egyptian 'shoal' seems something different from the generally accepted 'low-tide elevation', which is 'a naturally-formed area of land which is surrounded by and above water at low-tide but submerged at high tide' (Article 11, Paragraph 1 of the Geneva Convention).

With respect to the possibility of drawing straight baselines, the Decree of 1951 provides that 'where a shoal is situated not more than twelve nautical miles from the mainland or from an Egyptian island', the baselines are 'lines drawn from the mainland or the island and along the outer edge of the shoal' (Article 6c). 'Where an island is not more than twelve nautical miles from the mainland', the baselines are 'lines drawn from the mainland and along the outer shores of the island' (Article 6e). Detailed provisions of the Decree apply the 12-mile criterion to coastal and non-coastal groups of islands.[13]

It is evident that the Egyptian legislation of 1951 took into consideration the characteristics of the Red Sea (including the Gulf of Aqaba and the Gulf of Suez), which is dotted with numerous islands, shoals and reefs. If the above-mentioned provisions on islands are to be evaluated today, in the light of the subsequent evolution of the law of the sea, they appear to be unacceptable as regards the too broad definition of an island, but quite precise and moderate as regards the 12-mile criterion of distance. Unlike the Egyptian decree, the Geneva and Montego Bay conventions do not provide any geometrical guidance in order to determine where 'there is a fringe of islands along the coast in its immediate vicinity', and instances of excessive claims to draw straight baselines to very distant islands are not lacking in recent state practice.

It can be added that, according to generally accepted rules, straight baselines cannot be drawn to and from low-tide elevations, unless lighthouses or similar installations which are permanently above sea level have been built on them (Article 4, Paragraph 3 of the Geneva Convention).[14] The Egyptian decree is, on the contrary, very generous in granting a major role to shoals and in giving low-tide elevations the status of islands with respect to the drawing of straight baselines. However, the Egyptian legislation may be seen as a precedent for the recent provision which allows states having islands with fringing reefs to measure the territorial sea from the seaward low-water line of the reef (Article 6 of the Montego Bay Convention).

Despite its undeniable interest, the Egyptian legislation of 1951 remained for many years in a state of uncertainty. Egypt is not reported to have made publicly available any official maps or any lists of coordinates, which could have clarified the precise extent of what were claimed to be internal waters.

THE LEGISLATION OF 1990

By a recent Decree of the President of the Republic (No. 27 of January 9, 1990) Egypt established a system of straight baselines from which 'the maritime areas coming under the sovereignty and rule of the Arab Republic of Egypt, including its territorial sea, shall be measured'.[15] The points connected by the baselines are defined by coordinates, in accordance with the geodetic datum (Mercator projection), and are listed in two annexes, relating to the Mediterranean Sea and the Red Sea.

The preamble of the 1990 decree states that it was promulgated after having considered the 1951 decree.[16] The new legislation may thus have the merit of giving a more precise (though rather belated) implementation of some elastic provisions embodied in the old legislation.

As regards publicity, on August 26, 1983, upon ratification of the Montego Bay Convention, Egypt declared that it 'will publish at the earliest opportunity, charts showing the baselines from which the breadth of its territorial sea in the Mediterranean Sea and in the Red Sea are measured, as well as the lines marking the outer limit of the territorial sea, in accordance with usual practice'. Actually, Egypt deposited the 1990 decree with the United Nations Secretary-General in accordance with Article 16 of the Montego Bay Convention.[17] It does not seem however that Egypt has yet deposited the nautical charts showing the baselines as promised. Under Article 16 charts may be substituted by the deposit of the list of the geographical coordinates, specifying the geodetic datum.

The 1990 decree gives the coordinates of 53 points in the Mediterranean and 56 points in the Red Sea. Straight baselines are drawn along the whole extension of the Egyptian coastline, in the Mediterranean from the boundary with Libya to the boundary with Israel (527.1 nm in total) and in the Red Sea from the boundary with Israel to the so-called treaty-boundary with Sudan (525.6 nm in total; see the Appendix to this chapter). The longest segment is 40 nm and connects points 49 (a reef near Zagarbad island) and 50 (reef of Sha'ab Abu Fendera) in the Red Sea. The two shortest segments are both 0.9 nm long, between points 23 and 24 in the Mediterranean and between points 18 and 19 in the Red Sea. In general, the segments have a reasonable length.

As regards the Mediterranean, the Egyptian straight baselines give the rather paradoxical impression of being illegitimate though moderate. As the coast is devoid of relevant features, such as deep indentations or fringes of islands, there seems to be no legal basis for establishing a straight baseline system. Having however decided to establish such a system, Egypt took a rather moderate attitude in drawing the lines, which in many cases run very close to the shore (see, for instance, the eastern part of the coast).The result is that the straight baselines shift the limits of the Egyptian territorial sea and other coastal zones seaward very little (Figures 9.2 and 9.3).

In the case of bay-like indentations, Egypt avoided closing any of the above-mentioned gulfs of El Arab, Salum, Abu Hashaifa, Pelusium and El Arish, which do not comply with the semicircle rule. Straight baselines are instead drawn inside such gulfs. The closing line of the Gulf of Abu Qir is constituted by two segments (25–26 and 26–27) linking the two entrance points with the small island of Nelson, lying almost on the same axis of the headlands of the gulf. The sum of the lengths of the two segments does not exceed 24 nm, but there may be some doubts as to whether Abu Qir meets the semicircle criterion.

Also in the Gulf of Aqaba, the straight baselines closely follow the direction of a coastline which is neither deeply indented nor cut into. Some of the base points are located on reefs in the vicinity of the shore. The Egyptian straight baseline does not seem to affect in any substantial way navigation through the Gulf of Aqaba to and from the very short coasts of Israel and Jordan, enclosed at the end of the gulf. Such navigation is in any case protected by Article V, Paragraph 2 of the peace treaty between Egypt and Israel (Washington, March 26, 1979), according to which 'the parties consider the strait of Tiran and the Gulf of Aqaba to be international waterways open to all nations for unimpeded and non-suspendable freedom of navigation and overflight' (Figure 9.4).

The Gulf of Suez (Figure 9.5) is closed by a three-segment line across the strait of Gubal, linking successively Cape Muhammad, Shaker island (segment 33–34, 19.7 nm long), a reef or rock 4.7 nm east of Umm Qamar island (segment 34–35, 16 nm long), and Cape Abu Soma (segment 35–36, 20.2 nm long). While the Gulf of Suez does meet the semicircle rule (with respect to a circle having a diameter the sum of the lengths of the above-mentioned segments), the possibility of considering the so-called multi-mouthed bays as a series of bays in succession is still very questionable.

The enclosure of the Gulf of Suez by straight baselines should be carefully considered. In principle, it should not result in a prejudice to

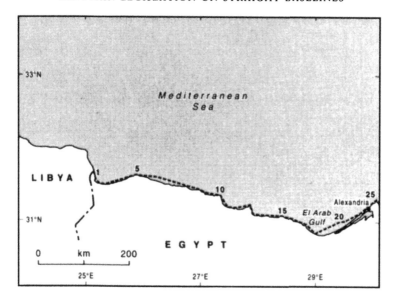

Figure 9.2 Egyptian straight baselines – western Mediterranean coast

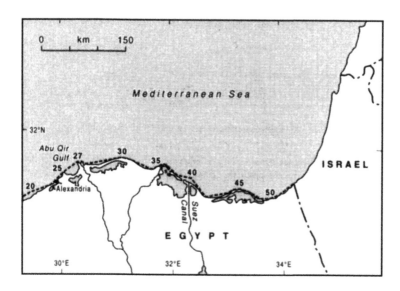

Figure 9.3 Egyptian straight baselines – eastern Mediterranean coast

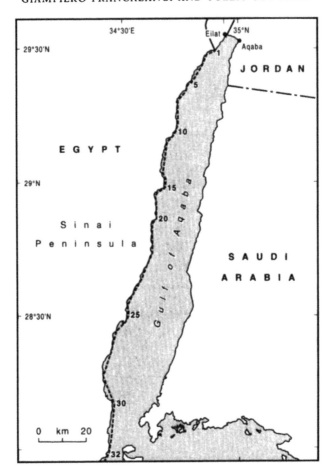

Figure 9.4 Egyptian straight baselines – Gulf of Aqaba

the freedom of navigation of any state using the Suez Canal (and, consequently, also to and from the canal), as provided in the multilateral convention signed in Constantinople on October 29, 1888. However, the application of the rule embodied in Article 5, Paragraph 2 of the Geneva Convention and in Article 8, Paragraph 2 of the Montego Bay Convention could lead to a regime of innocent passage through the Gulf of Suez.[18]

The remaining segments of the baseline in the Red Sea also follow in the majority of cases the general direction of the coast, linking basepoints which are often located on islands, shoals and reefs. In the

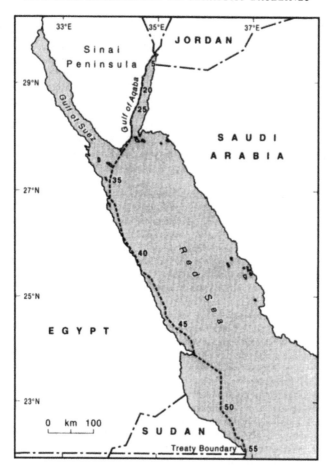

Figure 9.5 Egyptian straight baselines – Red Sea

southernmost part of the coast, from Cape Banas to Cape Adarba, the line seems however to depart to an appreciable extent from the general direction of the coast.

A full list of basepoints and distances is given in Appendix C on p. 138.

NOTES AND REFERENCES

1 This study is made within the framework of a research project on the technical aspects of the new law of the sea, sponsored by the Italian

Ministry of University and carried on at the University of Parma. G. Francalanci made the technical analysis and T. Scovazzi the legal comment.

2 On the straight baseline systems established by Egypt and by the other Mediterranean states see A. de Guttry, 'The delimitation of territorial waters in the Mediterranean', in *Syracuse Journal of International Law*, vol. 1, 1984, p. 377; T. Scovazzi, 'Bays and straight baselines in the Mediterranean', in *Ocean Development and International Law*, vol. 19, 1988, p. 401; A. Gioia, *Titoli storici e linee di base del mare territoriale*, Padova, 1990, p. 701. For the relevant maps see T. Scovazzi, G. Francalanci, S. Mongardini & D. Romano, *Atlas of the straight baselines*, 2nd edn. Milano, 1989.

3 Société des Nations, Comité d'experts pour la codification progressive du droit international, *Rapport au Conseil de la Société des Nations*, 1927, p. 257.

4 This may however be understood by pointing out that, in 1927, the legal distinction between the territorial sea and the internal or historic waters was still a rather confused issue. Moreover, there are a few states that still today claim a historic territorial sea (for instance Sri Lanka in the Gulf of Manar).

5 *Revue Egyptienne de Droit International*, 1952, p. 91.

6 Text in *Revue Egyptienne Droit International*, 1950, p. 175; in US Department of State, *Limits in the Seas*, No. 22, 1970; and in U. Leanza, L. Sico & M.C. Ciciriello, *Mediterranean Continental Shelf*, Dobbs Ferry, 1988, p. 239.

7 Text in ICJ, *Pleadings, Oral Arguments, Documents – Fisheries Case*, II, p. 260. A subsequent Royal Decree of Saudi Arabia of February 16, 1958 is analysed in US Department of State, *Limits in the Seas*, No. 20, 1970. For a similar Decree of Syria of December 28, 1963, see US Department of State, *Limits in the Seas*, No. 53, 1973.

8 Hereinafter: Geneva Convention.

9 Hereinafter: Montego Bay Convention.

10 'Territorial sea and the Continental Shelf – a review of recent developments', in *Revue Egyptienne de Droit International*, 1952, p. 127.

11 *Supra*, note 4. Also the United States protested against the 1951 Egyptian decree, see *Revue Egyptienne de Droit International*, 1951, p. 94.

12 The solution later followed by the drafters of the Geneva Convention was criticised in an Egyptian comment: 'The Egyptian bays are, in fact, well designated as "bays". The importance of these considerations is emphasised by the severity of the tests now proposed to be applied to "bays" both as to the area and breadth of opening – tests which by themselves would otherwise exclude from the definition most Egyptian and other Mediterranean "bays".' (Report of the International Law Commission on the regime of the High Seas and of the Territorial Sea with notes on the Egyptian position', in *Revue Egyptienne de Droit International*, 1955, p. 205).

13 'The following are established as the baselines from which the coastal sea of the Kingdom of Egypt is measured: ... f) where there is an island group which may be connected by lines not more than twelve nautical miles long, of which the island nearest to the mainland is not more than twelve nautical miles from the mainland, lines drawn from the mainland and along the outer shores of all the islands if the group of islands form a chain, or along the outer shores of the outermost islands of the group if the islands do not

form a chain; and g) where there is an island group which may be connected by lines not more than twelve nautical miles long, of which the island nearest to the mainland is more than twelve nautical miles from the mainland, lines drawn along the outer shores of all the islands if the group of islands form a chain, or along the outer shores of the outermost islands of the group if the islands do not form a chain' (Article 6).

14 Nevertheless, under Article 7, Paragraph 4 of the Montego Bay Convention straight baselines may be drawn to and from low-tide elevations also in instances where such drawing has received general international recognition.

15 Text in UN, Office for Ocean Affairs Law of the Sea, *Law of the Sea Bulletin*, No. 16, December 1990, p. 3. The present breadth of the Egyptian territorial sea is 12 nm as determined by a Decree of February 17, 1958. Egypt has a 24-mile contiguous zone and might also have an exclusive economic zone: in ratifying on August 26, 1983 the Montego Bay Convention it declared that:

> it will exercise as from this day the rights attributed to it by the provisions of parts V and VI of the United Nations Convention on the Law of the Sea in the exclusive economic zone situated beyond and adjacent to its territorial sea in the Mediterranean Sea and in the Red Sea.

It is however doubtful whether Egypt has so far implemented this declaration.

16 The preamble is however not reproduced in the text of the 1990 decree published in the *Bulletin* quoted above, note 15.

17 It might be noted that Egypt has complied with Article 16 before the entering into force of the Montego Bay Convention. This attitude deserves appreciation, taking into account that the objective of Article 16 is merely to give adequate publicity to what states have already enacted (or are going to enact) in their domestic legislation.

18 'Where the establishment of a straight baseline in accordance with Article 4 has the effect of enclosing as internal waters areas which previously had been considered as part of the territorial sea or of the high seas, a right of innocent passage, as provided in Articles 14 to 23, shall exist in those waters' (Article 5, Paragrah 2 of the Geneva Convention). Egypt is however not a party to the Geneva Convention.

APPENDIX C

ANALYSIS OF THE EGYPTIAN SYSTEM OF STRAIGHT BASELINES

According to Decree Decree No 27, 1990

Mediterranean Sea Coast   N.C. 3356, 2574, 2573 B.A.

No.	Point *(name or description)*	*Distance between points (nm)*
1	Land Boundary with Libya, Marsa el Ramla	
		6.5
2	Beacon Point, near Salum	
		4.8
3	Shoreline	
		4.7
4	Shoreline	
		29.9
5	Cape (?), Sidi Barrani	
		18.3
6	Taifa Rock, 3.2 nm N of the coast	
		21.0
7	Ishaila Rocks, 2 nm N of the coast	
		18.1
8	Rocks, Ras Abu Laho	
		4.9
9	Ras Umm El Rakham	
		15.0
10	Rocks, Ras Alam El Rum	
		11.5
11	Shoreline, Abu Ashafa Bay	
		8.3
12	Ras Abu Ashafa	
		11.9

13	Ras El Hekma	
14	Shoreline, near Fuka	9.1
15	Rad El Daba	26.3
16	Tannum Reef	8.7
17	Gibeisa Reef	13.3
18	El Shaqiq Reef	5.7
19	Shoreline, El Alamein (El Shamama Banks)	6.2
20	Victorieuse Rock	22.3
21	Abu Sir Reef	6.6
22	El Dikheila	15.7
23	Ras Eltin (lighthouse)	4.9
24	Ras Eltin	0.9
25	Shoreline, E of Abu Qir	10.9
26	Nelson Island (Disuqi Is.) N of Abu Qir Bay	4.0
27	Ras Umm El Nabayil, Rosetta Mouth	15.5
28	Ras Umm El Nabayil	1.3
29	Shoreline	5.3
30	Cape Burullus	30.1
31	Cape Burullus	4.4
32	Shoreline	3.9
33	El Aiyash, Fort no. 6	4.3
34	Shoreline, E of Fort Ashtum Gamasa	18.5
35	Lighthouse, W of Canal	8.4
36	Damietta Mouth	6.4
37	2.8 nm E of Damietta East lighthouse	1.8

		3.3
38	W side of Bay of Diba	
		12.4
39	Shoreline, Bay of Diba	
		12.2
40	W Mole, Suez Canal	
		18.3
41	Shoreline near El Farama	
		19.1
42	Mouth of Sebka El Bardawil	
		8.4
43	Shoreline near Katib El Qals	
		2.0
44	Closing sand dune, Sebka El Bardawil	
		2.3
45	idem	
		4.0
46	idem	
		6.3
47	idem	
		3.1
48	idem	
		8.2
49	Shoreline	
		9.8
50	Shoreline, W of El Arish	
		13.4
51	Shoreline	
		6.6
52	Shoreline	
		8.3
53	Land Boundary with Israel	
	Total length nm	527.1

Red Sea Coast   **N.C. 8, 63, 138 BA**

1	Land Boundary with Israel – Gulf of Aqaba	
		1.8
2	Shoreline	
		3.2
3	Shoreline	
		1.2
4	Shoreline	
		3.4
5	Shoreline	
		1.0
6	Shoreline	

APPENDIX C

		1.4
7	Shoreline	
		2.9
8	Shoreline	
		4.8
9	Ras El Burqa	
		2.0
10	Ras El Burqa	
		1.8
11	Shoreline	
		6.2
12	Reef (?)	
		2.7
13	Near the lighthouse, S of Nuweiba	
		1.7
14	Promontory, El Qarnus	
		1.5
15	Promontory, El Qarnus	
		1.1
16	Promontory, El Qarnus	
		1.6
17	Shore, Nuweiba El Muzeina	
		1.4
18	Shoreline	
		0.9
19	Shoreline	
		4.2
20	Cape (?)	
		1.3
21	Shoreline	
		6.8
22	Cape (?)	
		6.0
23	Ras Abu Galum	
		6.8
24	Shoreline	
		2.7
25	Lighthouse, Dahab	
		1.8
26	Reef	
		3.2
27	Shoreline	
		3.7
28	Cape (?)	
		6.8
29	Shoreline	
		6.9
30	Reef	
		6.7

141

31	Reef	
		4.8
32	Lighthouse, Ras Nusrani	
		18.2
33	Ras Muhammad Strait of Gubal	
		19.7
34	Lighthouse, Shaker Is. (Shadwan)	
		16.0
35	Reef or Rock, 4.7 nm E of Umm Qamar Is.	
		20.2
36	Ras Abu Soma, reef	
		6.8
37	Panorama Reef	
		3.6
38	Middle Reef	
		36.8
39	Lighthouse, Quseir	
		29.4
40	Reef, Ras Toronbi	
		13.6
41	Reef, Marsa Imbarak	
		13.0
42	Elphinstone Reef	
		37.6
43	Reef, 6.5 nm from shore	
		9.0
44	Wadi Gimal Island	
		16.0
45	Reef, Siyul Island	
		18.0
46	Fury Shoal	
		6.9
47	Fury Shoal	
		16.2
48	Ras Banas	
		36.8
49	Reef, 7.9 nm E–SE from Zagarbad Is.	
		40.0 (max.)
50	Reef, Sha'ab Abu Fendera	
		21.8
51	East edge, large shoal	
		16.3
52	Reef, large shoal	
		9.6
53	Reef	
		13.1
54	Ras Adarba	
		2.2

55	Shoreline, South of Ras Adarba	
		1.8
56	Land Boundary with Sudan, Lat 22°	

Total length nm   525.3

NB Distances have been measured for the Mediterranean sea on the Meridian at Latitude 30°, where 1' has a length of 1847 metres; for the Red Sea on the Meridian at mean Latitude 25° (1' = 1846 metres).

10

THE STABILITY OF LAND AND SEA BOUNDARY DELIMITATIONS IN INTERNATIONAL LAW

Geoffrey Marston

THE CONCEPT OF STABILITY

In reply to a recent Parliamentary question asking whether the UK Government would support a Kurdish State, the Prime Minister replied that it would not, because the UK was a party to the 1923 Treaty of Lausanne which had established the present frontiers of northern Iraq.[1] This affirmation of the continuity of obligations to respect a boundary reflects an aspect of the stability of boundary delimitations.

Two questions will be examined in this paper:

(i) Is the existence and alignment (*tracé*) of a line of delimitation established by treaty, or by third party determination pursuant to a treaty, binding on every party to the treaty until such time as it is altered, to use the words of the Helsinki Principles of 1975, 'in accordance with international law, by peaceful means and by agreement'? Or can one party, by invoking a valid ground for the treaty's invalidity or termination, or by taking counter-measures following an unlawful act by the other party, thereby regard itself as no longer legally obliged to respect the line? These questions relate to what will be called the line's internal stability, namely its applicability *inter partes*.

(ii) Is the existence and tracé of a line of delimitation established either by treaty or otherwise, binding on (or as international lawyers say opposable to) states other than those which established it? This question relates to what will be called the line's external stability, namely its opposability to other states, which could include – as in postwar settlements – the very states whose boundaries are thus delimited, and possibly its opposability to all states (*erga omnes*).

144

THE STABILITY OF LAND DELIMITATION

The word 'boundary' is appropriate to describe a line which delimits the contiguous land areas of states. R.Y. Jennings (currently President of the International Court of Justice (ICJ)) once remarked that '[a] boundary, which is a line, is different from a frontier, which properly speaking is a zone' (1967 121 *Académie de Droit International: Recueil des Cours*: 428). It should be noted, however, that in French the corresponding word is 'frontière'. Thus the French jurist Charles Rousseau wrote

Il est préférable d'admettre que la frontière, simple limitation territoriale, désigne non l'espace qui s'étend de chaque côté de la ligne de séparation politique, mais la ligne qui détermine où commencent et finissent les territoires relevant respectivement de deux Etats voisins.

(*Droit International Public*, vol. III, 1977: 232)

Internal stability of land delimitations

The following factors support the internal stability of land delimitations:

(i) Stability as a general rule of treaty law

The fundamental principle of treaty law known as *pacta sunt servanda* – that agreements ought to be honoured – itself implies that treaties contain a predisposition towards internal stability. Furthermore, Article 42(2) of the Vienna Convention on the Law of Treaties, 1969 (the 1969 Convention),[2] implies that treaties may be terminated only in accordance with strictly defined conditions. It reads in part: 'The termination of a treaty, its denunciation or the withdrawal of a party, may take place only as a result of the application of the provisions of the treaty or of the present Convention.'

(ii) The doctrine of executed treaty provisions

Once a boundary line has been established by treaty, whether or not it has been demarcated, its existence as a legal construction binding on the parties is no longer dependent on the continued existence of the treaty or treaty provision which established it. This is a consequence of the doctrine of executed treaty provisions. Sir Cecil Hurst, when legal adviser to the UK Foreign Office, remarked in 1922:[3]

145

A treaty which merely defines a frontier is spent from the moment when the treaty has come into force and both parties are pledged to the respect of that frontier. It is the frontier which is permanent, not the treaty in which it is laid down.

The doctrine is reflected in Article 70(1) of the 1969 Convention:

> Unless the treaty otherwise provides or the parties otherwise agree, the termination of a treaty under its provisions or in accordance with the present Convention:
>
> (b) does not affect any right, obligation or legal situation of the parties created through the execution of the treaty prior to its termination.

It is not proposed here to discuss whether execution itself always terminates the treaty. It is sufficient to note that it creates a legal situation – the boundary line – which one party cannot thereafter repudiate by a unilateral act. As the USA, Brazil, Argentina and Chile observed in a joint statement in 1960:

> It is a basic principle of international law that the unilateral will of one of the parties is not sufficient to invalidate a boundary treaty nor to liberate it from the obligations imposed therein. Only mutual agreement by both parties can modify its provisions or attribute competence to an international tribunal to judge questions which may arise regarding such an instrument.
>
> (M. Whiteman, *Digest of International Law*, vol. 3, 1964: 679)

(iii) Provisions in multilateral treaties

Article 62(2) of the 1969 Convention reads: 'A fundamental change of circumstances may not be invoked as a ground for terminating or withdrawing from a treaty: (a) if the treaty establishes a boundary'. The basis for this exception was stated in the commentary to its final draft articles on the law of treaties by the International Law Commission (ILC) to be as follows: 'treaties establishing a boundary should be recognised to be an exception to the rule because, otherwise the rule, instead of being an instrument of peaceful change, might become a source of dangerous frictions' (*Yearbook of ILC*, 1966, vol. II: 259).

It is likely that this exception is also a codification of existing customary law, as was stated by the ICJ in the *Fisheries Jurisdiction Case* (Jurisdiction) (ICJ Reports, 1974, at p. 18).

There is a similar formulation in Article 62(2) of the Vienna Convention on Treaties concluded between States and International Organisations and between Two or More International Organisations, 1986.[4]

The 1969 Convention also contains Article 56(4)(b) whereby a treaty not containing a provision regarding termination, denunciation or withdrawal is not subject to denunciation or withdrawal unless such a right 'may be implied by the nature of the treaty'. A boundary treaty is unlikely to raise such an implication. As the ICJ stated in the *Temple of Preah Vihear Case* (ICJ Reports 1962: 34), 'In general, when two countries establish a frontier between them, one of the primary objects is to achieve stability and finality.'

As already mentioned, Article 70(1)(b) of the 1969 Convention provides that a treaty's termination does not usually affect 'legal situations' created through prior execution of the treaty.

The following factors are hostile to the internal stability of land delimitations:

(i) Circumstances where the treaty is void

Where the treaty purporting to establish the boundary line is void *ab initio* or becomes void before the line is established, the line itself is a nullity. The 1969 Convention sets out the circumstances in which a treaty is void *ab initio*. These include fraud, corruption or coercion of a state representative, and coercion of a state by the threat or use of force. It is also possible for a treaty to be void by reason of the violation of provisions of a party's internal law regarding competence to conclude treaties (a matter debated in the course of the *Guinea-Bissau/Senegal Arbitration*) and, in certain circumstances, by reason of error in the treaty, but in these two instances Article 69(2)(b) provides that 'acts performed in good faith before the invalidity was invoked are not rendered unlawful by reason only of the invalidity of the treaty' – a provision which might be relevant to a boundary established by treaty.

Articles 53 and 64 provide respectively that a treaty which is in conflict with a 'peremptory norm of general international law' (*jus cogens*) is void, or in the case of a newly-emerging rule becomes void and terminates. In the case of the latter situation, Article 71(2)(b) provides that the treaty's termination

> does not affect any right, obligation or legal situation of the parties created through the execution of the treaty prior to its

termination, provided that those rights, obligations or situations may thereafter be maintained only to the extent that their maintenance is not in itself in conflict with the new peremptory norm of general international law.

Article 53 defines a peremptory norm as a norm 'accepted and recognised by the international community of States as a whole as a norm from which no derogation is permitted and which can be modified only by a subsequent norm of general international law having the same character', but the 1969 Convention does not identify any particular norm as being of this character. Leading the candidates for such norms likely to be relevant to boundary treaties are the principle of self-determination of peoples and the principle of permanent sovereignty over natural resources. The crucial question is whether a boundary established by treaty concluded against the wishes of a particular 'people' whose land was thus transferred from one state to another, could be regarded as a nullity at the instigation, for example, of a later government of the state making the transfer; likewise whether a boundary established by a treaty between parties one of which was 'unequal' to the other could similarly be destabilised. Despite an abundance of speculative writings, there has so far been no clear answer given by state practice or authoritative decision.

(ii) Counter-measures

International law permits a state which has suffered a wrong at the hands of another state to take non-forceful counter-measures proportionate to the wrong suffered (e.g. *US/France Air Services Award 1978; 54 International Law Reports*: 337–40). If one party to a boundary treaty refuses to respect the boundary, the other party has a choice; either to regard the boundary as still binding on both parties or, alternatively, itself to disregard the boundary. In the latter case, the legal consequence may well be that the boundary is abrogated by virtue of the counter-measure and not by virtue of the termination of the treaty through breach.

(iii) Circumstances where a boundary is inherently unstable

Where the tracé is defined by reference to locations which are in themselves unstable, such as the centre of a navigable channel, there will clearly often be a dispute as to the location of the boundary line. Here, however, it is not the existence of the boundary as a legal concept which is in doubt but only its tracé. Such situations, though frequently occurring, are not within the scope of this paper.

External stability of land boundaries

Where the land boundary is delimited unilaterally, this is not automatically opposable to any other state. One exception, in the context of decolonisation, is where the unilateral delimitation separated two administrative portions of the predecessor state's colonial territory. By virtue of the doctrine of *uti possidetis*, applied by the newly independent states in Latin America in the early 19th century, and adopted in principle for African boundaries by the Organisation of African Unity in 1964, the colonial line becomes the international boundary between the two successor states unless they agree otherwise. A Chamber of the ICJ in the *Case Concerning the Frontier Dispute: Burkina Faso/Mali*, in which both parties accepted the doctrine in advance, described it as 'a rule of general scope' (ICJ Reports 1986: 565). *Where the land boundary is delimited by treaty*, external stability may be incompatible with the general rule of treaty law sometimes called the 'third State' or *pacta tertiis* rule. This, as expressed in Article 34 of the 1969 Convention, reads: 'A treaty does not create either obligations or rights for a third State without its consent'. There is a widely-held opinion, however, that boundary treaties form an exception to this rule. Thus the French jurist Claude Blumann wrote, though with obvious surprise:

> Les accords frontaliers constituent des traités d'un type assez particulier. De l'avis général, ils dérogent au principe de l'effet relatif. Ils sont valables *erga omnes*. Cette situation a de quoi étonner, d'autant qu'il s'agit le plus souvent d'accords bilatéraux.
>
> (*La frontière* 1980: 12)

Article 38 of the 1969 Convention provides that the third state rule does not preclude a rule in a treaty from becoming binding upon a third state if it is also a recognised rule of customary international law. Are there any such rules? Four candidates come to mind:

(i) Objective regimes

It is often asserted that certain kinds of treaty produce 'objective regimes' opposable to all states and not just to the treaty parties. One example, endorsed by the ICJ in its Advisory Opinion on *Reparations for Injuries* (ICJ Reports 1949: 174) is the United Nations created by the Charter. However, the ICJ stressed the near-universality of membership of the Charter, a feature not present in the majority of boundary treaties. It is not denied that a treaty with only a few parties, or even a bilateral treaty, may affect the legal position of third states, but this is likely to occur, as in a peace settlement, where the parties intend the treaty to confer rights or obligations on third states and where third states are willing or prepared to accept such conferment, a situation recognised by Articles 35(1) and 36(1) of the 1969 Convention. These conditions are not likely to arise with bilateral boundary treaties in general.

(ii) Dispositive or 'real' treaties

It is sometimes asserted that a boundary is opposable to all states because it concerns land. As long ago as 1907, Berriedale Keith wrote:

> A boundary treaty, when completed, is not a contract but a conveyance, and the boundaries established are, as in the case of private law, good against the world. The cessionary or the conqueror cannot re-open the question on any legal grounds.
>
> (*The Theory of State Succession*: 27)

But as international law does not normally concern itself with land as such, but only with the spatial application of jurisdictional powers, the likening of a boundary treaty to a private law land transfer between individuals may be a false analogy.

(iii) State succession

In its report to the General Assembly in 1974, the ILC observed that the case for excluding boundaries from the 'clean slate' rule on a succession of states was 'strong and powerfully reinforced by the decision of the United Nations Conference on the Law of Treaties to except from the fundamental change of circumstances rule a treaty which establishes a boundary' (*Yearbook of ILC*, 1974, vol. II, part 1: 261). Article 16 of the Vienna Convention on Succession of States in respect of Treaties, 1978,[5] sets out the following 'clean slate' rule on decolonisation:

> A newly independent State is not bound to maintain in force, or to become a party to, any treaty only by reason of the fact that at the date of the succession of States the treaty was in force in respect of the territory to which the succession of States relates.

150

Article 11 of the same Convention provides that 'A succession of States does not as such affect (a) a boundary established by a treaty: or (b) obligations and rights established by a treaty and relating to the regime of a boundary.'

These provisions are drafted in a cautious way. The qualifications 'only by reason of the fact' in Article 16 and 'as such' in Article 11 indicate that other factors affecting a boundary on a succession of states are not precluded. The Chamber of the ICJ in the *Case concerning the Frontier Dispute: Burkina Faso/Mali*, however, considered that the exception in Article 11 was well established. It stated (ICJ Reports 1986: 566):

> There is no doubt that the obligation to respect pre-existing international frontiers in the event of a State succession derives from a general rule of international law, whether or not the rule is expressed in the formula *uti possidetis*.

State succession to land boundaries established by treaty does not provide a wide exception to the third state rule. A boundary which was not opposable to third states before succession does not become opposable to third states merely by virtue of the succession; in the absence of recognition by other states it will be binding only on the state or states which succeed to the territory of the former state, and, one may assume, it will remain opposable to the former colonial states which concluded the treaty.

(iv) The uti possidetis principle

Unless otherwise agreed between the newly independent states, on decolonisation the *uti possidetis* principle applies equally to a boundary established by the colonial state by treaty as it does to a boundary established unilaterally. In effect, like Article 11 of the 1978 Convention, the principle constitutes an exception to the 'clean slate' rule embodied in Article 16. It is doubtful whether the degree of external stability extends beyond the new state or states with the addition, on the basis of estoppel, of the original parties to the treaty; a third state will not be obliged to recognise the post-colonial boundary unless it had recognised the pre-colonial boundary.

In conclusion, the evidence for excluding boundary treaties from the third state rule is less than convincing. Perhaps the most that can be said about a boundary established by treaty is that an absence of protest by third states might lead more easily to a finding of tacit acceptance of the boundary on their part than would be the case with regard to the effects of some other kinds of treaty. Of course, it is possible for a boundary to be endowed with an increased chance of factual external stability if it is established by all the states likely to have claims in the area, as, for example, where a tripoint is established by a tripartite agreement.

THE STABILITY OF MARITIME DELIMITATIONS

Internal stability of maritime delimitations

Maritime delimitations differ from land delimitations in a number of factual ways: (a) they are usually geometrically constructed and rarely demarcated by fixed markers (though see UK/Venezuela Treaty, 1942 in respect of the Gulf of Paria); (b) the administration of human settlement, so important in respect of a disputed land boundary, is not relevant to them; (c) the treaties establishing them are usually bilateral and between the states whose respective maritime areas are to be delimited, whereas some important land boundaries have been established by multilateral treaties sometimes without the participation of affected territorial states; (d) their tracés may be subject to a greater risk of instability through changeable reference points, particularly the low water line.

Maritime delimitations differ from land delimitations in a number of juridical ways: (a) Some lines of maritime delimitation are in law merely lines of allocation which, to use Shalowitz's words 'are not true boundaries ... but represent a dividing line for the inclusion or exclusion of certain land and island areas under a particular agreement' (*Shore and Sea Boundaries*, vol. 2, 1964: 385); (b) the maritime area is not jurisdictionally homogeneous and contains areas under coastal state sovereignty (internal waters and territorial sea) and areas under a less intensive regime (contiguous zone, exclusive economic or fisheries zone, continental shelf); (c) it might be said that they delimit areas which have no juridical existence apart from the existence of the land, or rather coastal front, to which they are appurtenant; (d) there are rules for maritime delimitation provided for in general multilateral treaties, whereas there are no such provisions in respect of land boundaries; (e) states have a discretion in the breadth of the maritime zones they claim, up to the limits laid down by customary or treaty law, the exercise of which in certain circumstances could produce a potentially destabilising situation.

The question is whether these and other distinctions give to maritime delimitations a quality of stability different from that of land delimitations discussed above.

The following factors support the internal stability of maritime delimitations:

(i) None of the above points of distinction would seem sufficient for excluding lines of maritime delimitation from the effect of the doctrine of executed treaty provisions, a doctrine which applies to a wide variety of legal situations not confined to boundaries.

(ii) The rule of fundamental change of circumstances, excluded from being a destabilising factor in respect of 'boundaries', as evidenced by Article 62(2)(a) of the 1969 Convention, was considered by the ICJ in the *Aegean Continental Shelf Case* to be excluded also in respect of maritime delimitations. It stated (ICJ Reports 1978: 35–6):

> Whether it is a land frontier or a boundary line in the continental shelf that is in question, the process is essentially the same, and inevitably involves the same element of stability and permanence, and is subject to the rule excluding boundary agreements from fundamental change of circumstances.

This observation was regarded by a majority of the arbitrators in the *Guinea-Bissau/Senegal Arbitration* in 1989 (83 *International Law Reports*: 36) as constituting a precedent for not drawing any distinction between land and maritime delimitations for the purpose of applying the *uti possidetis* rule.

(iii) Both before and after the conclusion of the 1969 Convention, the word 'boundary' (French 'frontière') has been used in many treaties concluded by a wide variety of states to describe the line of delimitation not only between internal waters or territorial seas but also between zones in which the coastal states have a less intensive legal regime. If states had thought that the word 'boundary' as used in the 1969 and 1978 Conventions was confined to land delimitation lines they would have taken care to use other expressions for maritime delimitations such as 'line of delimitation' or 'limits'. The indications are that states are not conscious of any need to apply a different terminology to maritime delimitations. As the French jurist Prosper Weil observed, 'The term "boundary", used more and more frequently today in place of "delimitation line", is not, as one can see, a usurpation' (Weil 1989: 94).

The following factors are hostile to the internal stability of maritime delimitations:

(i) The causes of treaty invalidity producing nullity of a boundary line should apply equally to treaties purporting to establish maritime delimitations.

(ii) The discussion above of non-forcible counter-measures should apply equally to maritime delimitations.

(iii) There is an additional potentially destabilising argument, however. The principle – stated more than once by the ICJ – that the land dominates the sea has two significant corollaries: (1) The maritime zones under coastal state jurisdiction have no legal existence in the absence of a coastal front, just as a shadow cannot exist independently of the object which creates it; (2) it is the coastal front of a state which generates its legal entitlement to maritime zones, whether the zone is automatically appurtenant (such as the continental shelf and the bed and subsoil of the territorial sea) or has to be positively claimed (such as the contiguous zone and the exclusive economic zone). These two corollaries lead to another necessary conclusion, namely that every notional point in a state's maritime area owes its existence to a notional point on its coastal front, which can only be the point to which it is nearest. Where coastal states are adjacent or opposite to each other this linkage produces an ideal or platonic delimitation line, namely an equidistance or median line, even though there has not yet been any agreement to delimit. Delimitation, when it occurs, is thus merely declaratory, as the ICJ observed in the *North Sea Continental Shelf Cases* (ICJ Reports 1969: 22 (paras. 18–20)). Should states agree to construct a delimitation line which departs from the platonic line it would amount to a purported transfer to one state of maritime space indissolubly linked to the coastal front of the other state, a legal impossibility and thus a nullity. Hence the relevance of the remark of R.Y. Jennings that maritime delimitation concerns the boundaries 'not of alienable territory but of inalienable appurtenances of territory' (*The Acquisition of Territory in International Law* 1963: 14–15, footnote).

The argument would then go on to point out that the present theory of maritime zones, evolved in particular by the ICJ in the *Libya/Malta Case* (ICJ Reports 1985: 13), which regards simple distance as the criterion of entitlement for all zones within 200 nm is more propitious to a mathematical concept of appurtenance than the 'natural prolongation' of the *North Sea Continental Shelf Cases*. As M. Bedjaoui remarked in his dissenting opinion in the *Guinea-Bissau/Senegal Arbitration* in 1989: '[o]ne of the great innovations in the contemporary law of the sea is that it recognises a right to a maritime territory which exists independently of, and prior to, any delimitation' (83 *International Law Reports*: 49–50).

It is evident, however, that this argument does not reflect the real world of maritime delimitation. The 'fundamental norm' of delimitation in customary law, namely that 'delimitation, whether effected by direct agreement or by the decision of a third party, must be based on the application of equitable criteria and the use of practical methods capable of ensuring an equitable result' (*Gulf of Maine Case* ICJ Reports 1984: 299–300) necessarily implies that there is no unique platonic line of delimitation, let alone one based on equidistance, waiting to be declared. The *Libya/Malta Case*, rather than endorsing a concept of mathematical inevitability for equidistance, denied that the method of equidistance was obligatory even as a preliminary step. Turning to multilateral treaties, Article 12 of the Geneva Convention on the Territorial Sea and the Contiguous Zone, 1958, in its Articles 12 and 24(3), implies that states may validly make delimitation agreements for the territorial sea and contiguous zone respectively which do not comply with equidistance. A similar implication is found in Article 6(1) and (2) of the Geneva Convention on the Continental Shelf, 1958. It could be argued, however, that as the customary rule of equitable result is stated to apply equally to agreements, an inequitable delimitation established by a treaty is a nullity. But this would overlook the clear principle of international law as expressed by the ICJ in the *North Sea Continental Shelf Cases* that 'it is well understood that, in practice, rules of international law can, by agreement, be derogated from in particular cases, or as between particular parties' (ICJ Reports 1969: 42). Furthermore, there is no evidence that the requirement of an equitable result is a 'peremptory norm'. The majority in the *Guinea-Bissau/Senegal Arbitration* considered that there was no rule of international law by which the equitableness of a line of maritime delimitation could later be verified or reviewed (83 *International Law Reports*: 42–3).

As regards state practice evidenced in maritime delimitation treaties, there are several examples of delimitation lines which are explicable only on the basis that the parties regarded themselves as free to negotiate whatever tracé was appropriate to their particular requirements and bargaining strengths. Thus the Canada/France Fisheries Agreement of March 27, 1972 provides that the line of maritime delimitation between the coasts of Canada and of the French islands of St Pierre and Miquelon runs along the low water line on the Canadian islets of Little Green Island and Enfant Perdu. This deprives Canada of the maritime zones otherwise generated by those coasts. Likewise, the then legal adviser to the French Ministry of Foreign Affairs, M. Guillaume (now a judge of the ICJ), wrote that the maritime delimitation agreement between France and Monaco of 1984 'was inspired by considerations of courtesy and good-neighbourliness, and adopted an *ad hoc* solution which had nothing to do with law and is explained only by the special nature of the relations between the two countries' (translation in Weil 1989: 112, footnote 32).

Weil summed up the current position as follows (Weil 1989: 112; (see to a similar effect, Bardonnet 1989: 3)):

> If negotiated delimitations were subjected to legal norms, this would mean that there were *jus cogens* rules restricting the contractual freedom of States and, as a corollary, that a delimitation agreement disregarding such rules would arguably be invalid. This is certainly not the case. The principles and rules of international law set out by the courts in respect of maritime delimitation are purely suppletive and States are free to agree to reject them.

As between parties to the 1982 Convention on the Law of the Sea, the application to delimitation agreements of the requirement of an 'equitable solution' found in Article 74(1) (exclusive economic zone) and Article 83(1) (continental shelf) might raise an argument, at least before the future International Tribunal to be established by the Convention, that a line of delimitation which produced an inequitable solution, if the parties so agreed to establish it, was a nullity, being contrary to one of the Convention's 'basic principles' and thus violating Article 311(3) of the Convention. As the Convention is not yet in force, and the Tribunal not yet in place, such an argument is currently hypothetical.

(iv) In its discussion on the text leading to the Vienna Convention on Treaties concluded between States and International Organisations or between Two or More International Organisations, 1986, the ILC observed that in its view Article 62 of the 1969 Convention 'may not extend to certain lines of maritime delimitation, even if, in all intents and purposes, they constitute true boundaries' (*Yearbook of ILC*, 1980, vol. II, part I: 81). However, as discussed above, the doctrine of executed treaty provisions would usually maintain the legal concept of the boundary in existence notwithstanding the disappearance for any reason of the treaty which established it.

External stability of maritime delimitations

Where the maritime delimitation is made unilaterally, as the ICJ stated in the *Anglo-Norwegian Fisheries Case* (ICJ Reports 1951: 132), a maritime delimitation 'cannot be dependent merely on the will of the coastal State ... the validity of the delimitation with regard to other States depends upon international law'.

On a decolonisation, it is otiose to ask whether the successor State is obliged to respect the claim made by its colonial predecessor, since if the claim is in accordance with international law the new State may, if it wishes, adopt the delimitation as its own, or, if the claim is contrary to international law, it cannot be used against any other State, except, perhaps, against the colonial predecessor on the basis of an estoppel.

In the rare situation of a maritime delimitation by a colonial state delimiting two administrative units of its territory, there would be considerable argument whether the doctrine of *uti possidetis* would apply so as to make the line binding on the successor states if they should fail to agree otherwise. It will be noted below that this issue arose in the *Guinea-Bissau/Senegal Arbitration* where the alleged delimitation line had been established by treaty. Quite apart from the *uti possidetis* argument, however, successor states become independent within the same area of spatial competence claimed by their predecessor state, in which case the unilateral colonial line would serve the same function in this respect as a land boundary, binding both states in the absence of agreement to change it.

Where the maritime delimitation is made by treaty, although external stability may be increased if all states with claims in a certain maritime area agree to a particular delimitation, for example, a tripartite treaty establishing a tripoint, the 'third State' rule makes maritime delimitations established by treaty inopposable to non-parties. Thus in the *North Sea Continental Shelf Cases*, the line established by treaty between Denmark and the Netherlands on March 31, 1960 was held not to be opposable to the Federal Republic of Germany, a non-party to it. In the *Case Concerning the Frontier Dispute (Burkina Faso/Mali)*, the Chamber of the ICJ stated:

> in continental shelf delimitations, an agreement between the parties which is perfectly valid and binding on the treaty level may, when the relations between the parties and a third State are taken into consideration, prove to be contrary to the rules of international law governing the continental shelf.
>
> (ICJ Reports 1985: 578)

In the discussion on land delimitations above, it was concluded that the doctrines of state succession to treaties and of *uti possidetis* provide exceptions, though not far-reaching exceptions, to the third state rule. Do these doctrines apply at all to maritime delimitations?

(i) State succession

If the doctrine of executed treaty provisions applies to lines of maritime delimitation, and there seems to be no good reason why it should not, then the successor state or states are obliged, in the absence of agreement to do otherwise, to respect the line, since it defines the area of spatial competence appurtenant to the land territories to which they have respectively succeeded. There would thus be no need to consider whether such a line is or is not a 'boundary' for the purposes of an exception to the 'clean slate' rule.

(ii) Uti possidetis

The question of the application of *uti possidetis* to maritime delimitations established by treaty in the colonial period has arisen in two recent arbitrations:

(1) *Guinea-Bissau/Guinea*, 1985 (77 *International Law Reports* 635): The tribunal held unanimously that the 1886 France/Portugal treaty had not established a maritime delimitation at all but was merely a line allocating to one state or the other land territory situated on each side of the line. It was thus unnecessary to consider whether the doctrine of *uti possidetis* could have applied if it had been a true maritime delimitation.

(2) *Guinea-Bissau/Senegal*, 1989 (83 *International Law Reports* 1): Here the tribunal had to consider the application to the two states of a line established by treaty between Portugal and France in 1960 and purportedly delimiting the territorial sea, contiguous zone and continental shelf. A division of opinion resulted. The majority (MM. Barberis and Gros) considered that by their conduct the two successor states had accepted *uti possidetis* as applicable to the delimitation, and, furthermore, there was some state practice elsewhere supporting its application to maritime delimitations. M. Bedjaoui, dissenting, considered that the principle was inapplicable for a number of reasons: maritime delimitations were not within the purview of the Organisation of African Unity in 1964; the doctrine applied only to lines delimiting sovereignty and not areas of lesser jurisdiction; the zones beyond the territorial sea

were not 'territory'; the doctrine was designed to give stability in respect of human settlement and was not to be extended to situations where this was not appropriate; the state practice cited by the majority did not on investigation support their conclusion. To these reasons given by M. Bedjaoui might be added a further one, at least in respect of the continental shelf and exclusive economic zone, namely that automatically to apply *uti possidetis* would be incompatible with the customary law requirement of an 'equitable result'. But even if M. Bedjaoui is right and the *uti possidetis* doctrine is not applicable to maritime delimitations as a general rule, the doctrine of executed treaty provisions will still provide a basis for the continued validity of the line in delimiting the spatial areas to which each of the new states has succeeded.

CONCLUSION

In respect of land boundaries, the rules of international law provide a considerable measure of internal stability derived in particular from the fundamental principle of *pacta sunt servanda* and the doctrine of executed treaty provisions, supported by the exclusion of boundary treaties from the destabilising effects of a fundamental change of circumstances. Although there is a widely-held opinion that boundaries are opposable *erga omnes* this is difficult to reconcile with the 'third party' rule. There is, however, a limited degree of formal external stability, opposable only to successor states, provided by the exclusion of boundaries from the 'clean slate' rule of state succession and by the operation of the *uti possidetis* principle on a decolonisation.

In respect of lines of maritime delimitation, there does not appear to be any difference in legal principle between their internal stability and that of land boundaries. In particular, the newly emerged concept of equitable result would seem not to have derogated from the purely voluntarist nature of treaties establishing maritime delimitations. The scope for the application of rules of *jus cogens*, though perhaps not excluded entirely from future development, is restricted. The external stability of maritime delimitations would not seem to differ in principle from that of land delimitations, even if there remains doubt over the application to the maritime area of *uti possidetis*. To quote again Prosper Weil (Weil 1989: 94; see to a similar effect, Bardonnet 1989: 64), whereas 'the process of maritime delimitation is and remains an exercise *sui generis*; the dividing line to which it leads is undoubtedly very like a land boundary'.

NOTES

1 Parliamentary Debates, House of Commons, 6th series, vol. 184, Written Answers, col. 287, January 24, 1991.
2 In force from January 27, 1980: it currently has about sixty parties including the UK.
3 Public Record Office, London, reference FO 371/11737, f.201v [file N 1146/ 43/97]. See also a similar view expressed by Iraq to the Security Council in 1969 concerning the Shatt-al-Arab boundary (S/9185 and S/9323).
4 Not yet in force.
5 Not yet in force.

REFERENCES

Bardonnet, D. (1989) 'Frontières terrestres et frontières maritimes', *Annuaire français de droit international 1989*, 1–64.
Jennings, R.Y. (1963) *The Acquisition of Territory in International Law*, Manchester: Manchester University Press.
Kaikobad, K.H. (1983) 'Some observations on the doctrines of continuity and finality of boundaries', *British Year Book of International Law 1983*, 54, 119–41.
La frontière (1980) Colloque de la Société française pour le droit international (Poitiers, 1979), Paris: Pedone.
Shalowitz, A.L. (1964) *Shore and Sea Boundaries*, Washington: US Government Printing Office.
Weil, P. (1989) *The Law of Maritime Delimitations – Reflections*, Cambridge: Grotius Publications (translated by M. MacGlashan from *Perspectives du droit de la délimitation maritime*, Paris: Pedone, 1988).

ANNEX 1

The geographic positions set forth in this Annex are based on the World Geodetic System 1984 ('WGS 84') and, except where noted, are connected by geodetic lines. One nautical mile equals 1,852 meters. The maritime boundary is defined as follows:

From the initial point, 65° 30′ N., 168° 58′ 37″ W., the maritime boundary extends north along the 168° 58′ 37″ W. meridian through the Bering Strait and Chukchi Sea into the Arctic Ocean as far as permitted under international law.

From the same initial point, the maritime boundary extends southwestward connecting the following geographic positions:

2. 65° 19′ 58″ N., 169° 21′ 38″ W.
3. 65° 09′ 51″ N., 169° 44′ 34″ W.
4. 64° 59′ 41″ N., 170° 07′ 23″ W.
5. 64° 49′ 26″ N., 170° 30′ 06″ W.
6. 64° 39′ 08″ N., 170° 52′ 43″ W.
7. 64° 28′ 46″ N., 171° 15′ 14″ W.
8. 64° 18′ 20″ N., 171° 37′ 40″ W.
9. 64° 07′ 50″ N., 172° 00′ 00″ W.
10. 63° 59′ 27″ N., 172° 18′ 39″ W.

11. 63° 51′ 01″ N., 172° 37′ 13″ W.
12. 63° 42′ 33″ N., 172° 55′ 42″ W.
13. 63° 34′ 01″ N., 173° 14′ 07″ W.
14. 63° 25′ 27″ N., 173° 32′ 27″ W.
15. 63° 16′ 50″ N., 173° 50′ 42″ W.
16. 63° 08′ 11″ N., 174° 08′ 52″ W.
17. 62° 59′ 29″ N., 174° 26′ 58″ W.
18. 62° 50′ 44″ N., 174° 44′ 59″ W.
19. 62° 41′ 56″ N., 175° 02′ 56″ W.
20. 62° 33′ 06″ N., 175° 20′ 48″ W.

21. 62° 24′ 13″ N., 175° 38′ 36″ W.
22. 62° 15′ 17″ N., 175° 56′ 19″ W.
23. 62° 06′ 19″ N., 176° 13′ 59″ W.
24. 61° 57′ 18″ N., 176° 31′ 34″ W.
25. 61° 48′ 14″ N., 176° 49′ 04″ W.
26. 61° 39′ 08″ N., 177° 06′ 31″ W.
27. 61° 29′ 59″ N., 177° 23′ 53″ W.
28. 61° 20′ 47″ N., 177° 41′ 11″ W.
29. 61° 11′ 33″ N., 177° 58′ 26″ W.
30. 61° 02′ 17″ N., 178° 15′ 36″ W.

31. 60° 52′ 57″ N., 178° 32′ 42″ W.
32. 60° 43′ 35″ N., 178° 49′ 45″ W.
33. 60° 34′ 11″ N., 179° 06′ 44″ W.
34. 60° 24′ 44″ N., 179° 23′ 38″ W.
35. 60° 15′ 14″ N., 179° 40′ 30″ W.,
36. 60° 11′ 39″ N., 179° 46′ 49″ W.;

thence, it extends along an arc with a radius of 200 nautical miles and a center at 60° 38′ 23″ N., 173° 06′ 54″ W. to,

37. 59° 58′ 22″ N., 179° 40′ 55″ W.;

thence, it extends southwestward along the rhumb line, defined by the following points: 64° 05′ 08″ N., 172° 00′ 00″ W., 53° 43′ 42″ N., 170° 18′ 31″ E. to,

38. 58° 57′ 18″ N., 178° 33′ 59″ E.;

thence, it extends along an arc with a radius of 200 nautical miles and a center at 62° 16′ 09″ N., 179° 05′ 34″ E. to,

39. 58° 58′ 14″ N., 178° 15′ 05″ E.;
40. 58° 57′ 58″ N., 178° 14′ 37″ E.

41. 58° 48′ 06″ N., 177° 58′ 14″ E.
42. 58° 38′ 12″ N., 177° 41′ 53″ E.
43. 58° 28′ 16″ N., 177° 25′ 34″ E.
44. 58° 18′ 17″ N., 177° 09′ 18″ E.
45. 58° 08′ 15″ N., 176° 53′ 04″ E.
46. 57° 58′ 11″ N., 176° 36′ 52″ E.
47. 57° 48′ 04″ N., 176° 20′ 43″ E.
48. 57° 37′ 54″ N., 176° 04′ 35″ E.
49. 57° 27′ 42″ N., 175° 48′ 31″ E.
50. 57° 17′ 28″ N., 175° 32′ 28″ E.

51. 57° 07′ 11″ N., 175° 16′ 27″ E.
52. 56° 56′ 51″ N., 175° 00′ 29″ E.
53. 56° 46′ 29″ N., 174° 44′ 32″ E.
54. 56° 36′ 04″ N., 174° 28′ 38″ E.
55. 56° 25′ 37″ N., 174° 12′ 46″ E.
56. 56° 15′ 07″ N., 173° 56′ 56″ E.
57. 56° 04′ 34″ N., 173° 41′ 08″ E.
58. 55° 53′ 59″ N., 173° 25′ 22″ E.
59. 55° 43′ 22″ N., 173° 09′ 37″ E.
60. 55° 32′ 42″ N., 172° 53′ 55″ E.

61. 55° 21′ 59″ N., 172° 38′ 14″ E.
62. 55° 11′ 14″ N., 172° 22′ 36″ E.
63. 55° 00′ 26″ N., 172° 06′ 59″ E.
64. 54° 49′ 36″ N., 171° 51′ 24″ E.
65. 54° 38′ 43″ N., 171° 35′ 51″ E.
66. 54° 27′ 48″ N., 171° 20′ 20″ E.
67. 54° 16′ 50″ N., 171° 04′ 50″ E.

68. 54° 05' 50" N., 170° 49' 22" E.
69. 53° 54' 47" N., 170° 33' 56" E.
70. 53° 43' 42" N., 170° 18' 31" E.
71. 53° 32' 46" N., 170° 05' 29" E.
72. 53° 21' 48" N., 169° 52' 32" E.
73. 53° 10' 49" N., 169° 39' 40" E.
74. 52° 59' 48" N., 169° 26' 53" E.
75. 52° 48' 46" N., 169° 14' 12" E.
76. 52° 37' 43" N., 169° 01' 36" E.
77. 52° 26' 38" N., 168° 49' 05" E.
78. 52° 15' 31" N., 168° 36' 39" E.
79. 52° 04' 23" N., 168° 24' 17" E.
80. 51° 53' 14" N., 168° 12' 01" E.

81. 51° 42' 03" N., 167° 59' 49" E.
82. 51° 30' 51" N., 167° 47' 42" E.
83. 51° 19' 37" N., 167° 35' 40" E.;
84. 51° 11' 22" N., 167° 26' 52" E.;

thence, it extends along an arc with a radius of 200 nautical miles and a center at 54° 29' 42" N., 168° 05' 25" E. to,

85. 51° 12' 17" N., 167° 15' 35" E.;
86. 51° 09' 09" N., 167° 12' 00" E.
87. 50° 58' 39" N., 167° 00' 00" E.

DEPARTMENT OF STATE
WASHINGTON

June 1, 1990

Excellency:

I have the honor to refer to the Agreement between the United States of America and the Union of Soviet Socialist Republics on the Maritime Boundary, which has been signed by representatives of our two Governments today. I have the further honor to propose that, pending the entry into force of that Agreement, the two Governments agree to abide by the terms of that Agreement as of June 15, 1990.

On the basis of the foregoing, I have the honor to propose to Your Excellency that if the terms stipulated herein are acceptable to the Government of the Union of Soviet Socialist Republics, this note and Your Excellency's reply shall constitute an agreement between the two Governments, which shall enter into force on the day of your reply.

I avail myself of this opportunity to renew to Your Excellency the assurances of my highest consideration.

James W. Baker III

His Excellency
    Eduard A. Shevardnadze,
        Minister of Foreign Affairs of the
            Union of Soviet Socialist Republics.

ANNEX 1

LS NO.   132287
BL
Russian

THE MINISTER OF FOREIGN AFFAIRS
OF THE UNION OF SOVIET SOCIALIST REPUBLICS

Washington, June 1, 1990

Dear Mr. Secretary:

I have the honor to confirm receipt of your note of June 1, 1990 with the following content:

[text of letter from Secretary of State James Baker III to Soviet Foreign Minister Eduard A. Shevardnadze, showing no discrepancies in the two languages]

I have the honor to inform you that the Soviet side expresses its agreement that this note and and Your Excellency's reply shall consitute an agreement between the two Governments which shall enter into force on this day.

I avail myself of this occasion to renew to Your Excellency the assurances of my highest consideration.

[signed] E. Shevardnadze

CERTIFICATION OF TRANSLATION

I hereby certify that the attached translation bearing LS No. 132287 was prepared by the Office of Language Services of the Department of State and that it is a correct translation to the best of my knowledge and belief.

Dated: June 6, 1990

Chief, Translating Division

165

ANNEX 2

1977 EXCHANGE OF NOTES
BETWEEN THE UNITED STATES AND U.S.S.R.

United States Note No. 99 to the Soviet Government
January 24, 1977

(Complimentary opening) . . . and has the honor to refer
to the Agreement between the Government of the United States
of America and the Government of the Union of Soviet Socialist
Republics concerning fisheries off the coast of the United
States, signed in Washington, November 26, 1976, and to the
Fishery Conservation and Management Act of 1976. The
Embassy has the further honor to refer to the Decree of the
Presidium of the Supreme Soviet of the Union of Soviet Socialist
Republics of December 10, 1976 regarding fisheries.

The Embassy notes that the Fishery Conservation and
Management Act establishes, effective March 1, 1977, a
fishery conservation zone off the coast of the United States
extending 200 nautical miles seaward from the baseline from
which the territorial sea is measured, but it does not purport
to delimit the zone with adjacent or opposite states. However,
for practical reasons the Government of the United States
must determine prior to March 1, 1977, the limits to which it
will enforce its fishery jurisdiction.

The Embassy wishes to assure the Government of the
Union of Soviet Socialist Republics that in exercising its
fishery jurisdiction the Government of the United States of
America intends to act with full regard for treaties between
the two countries. Accordingly, the United States Government,
in enforcing its fishery jurisdiction, intends to respect the line
set forth in the Convention, signed at Washington March 30,
1867. The Government of the United States of course
anticipates that the Government of the Union of Soviet
Socialist Republics will follow a similar practice in exercising
its fishery jurisdiction under the Decree of December 10, 1976.
(Complimentary closing.)

U.S.S.R. Note No. 11/dusa to the United States, February 24, 1977

(Complimentary opening) . . . and refers to Note No. 99 of the U.S. Embassy dated January 24, 1977, and the honor to relate the following: the Government of the Union of Soviet Socialist Republics has taken into account the intention of the United States side, in setting forth its fisheries jurisdiction, to respect the Line established by the Convention signed April 18 (30) 1867 in Washington, D.C. The Government of the Union of Soviet Socialist Republics in carrying out its measures ensuing from the Decree of the Presidium of the Supreme Soviet of December 10, 1976, Temporary Measures for the Protection of Living Resources and Regulation of Fisheries in Areas Adjacent to the Union of Soviet Socialist Republics coast, "intends to adhere to the same line of the Convention of April 18, 1867, in the Arctic Ocean, Chukchi, and the Bering Seas.

The Government of the Union of Soviet Socialist Republics proceeds on the basis that the Government of the United States agrees that the line established by the 1867 Convention will be used to fix the limits of fishing zones for each of the two countries in those maritime areas being 200 miles beyond the coast of the United States and the Union of Soviet Socialist Republics, including the islands belonging to them."

U.S.S.R. Note No. 17/dusa to the United States, April 8, 1977

To clarify its reference to the 1867 Convention mentioned in the above note, the U.S.S.R. Ministry of Foreign Affairs presented the following Note to the U.S. Embassy in Moscow on April 8, 1977:

(Complimentary opening) . . . and referring to its Note No. 11/dusa of February 24, 1977, has the honor to confirm that this Note deals with the very same Russian-American Convention signed in Washington on March 18 (30), 1867, which is mentioned in the Embassy's Note No. 99 of January 24, 1977.

INDEX

168

Printed and bound by CPI Group (UK) Ltd, Croydon, CR0 4YY

01/11/2024

01782616-0004